Legros on Tanks and Traction in WW1

Pioneers of Armour Special 1

By Andrew Hills

Cover image Clayton 110 h.p. tractor climbing a 40° gradient with the front wheel off the ground; fitted with an inclinometer to show the steepness of the slope. Colourisation by Jaycee Davis and composition by Edward Jackson.

Copyright © FWD Publishing 2021
All Rights Reserved
ISBN: 9798680043066

Contents

	Page
Contents	i
Foreword	iv
Publisher's Note	viii
Traction on Bad Roads or Land	**1**
Part I	**3**
Four-Wheel Drive	4
Four-Wheel Steering	4
Four-Wheel Braking	5
Four-Wheel Driving	7
Automatic Differential Locking Gears	7
Four-Wheel Drive, Front Steering	11
Four-Wheel Drive, Four-Wheel Steering	12
Four-Wheel Independent Drive with Four-Wheel Steering	18
Applications of Four-Wheel Drive to Bad Roads or Land	21
Conclusions	21
Part II	**22**
Chain-Track Tractors	23
Classification of Chain-Track Tractors	26
Methods of Springing	28
Truck-Frames and Track-Frame Connexions	30
Steering	31
Driving Gear	32
Chain-Tracks	32
Track Shoes	37
Engines	38
Ignition and Starting	39
Radiators	39
Speed of Tractors	40
Drawbar Pull	40
Climbing Power	41
Drawbar Connexion	42
Leading Features of Tractors	42
The Log-Hauler, Phoenix	42
The Allis-Chalmers Tractor-Truck	44
The Caterpillar, Holt	44
The Clayton Tractor	50

The Tracklayer, C. L. Best	54
The Creeping-Grip Tractor, Bullock	56
The Austin Tractors	57
Burford-Cleveland Tractor	57
The Strait Tractor, Killen-Strait	58
The Ball-Tread Tractor, Yuba	59
Martin's Agricultural Tractor	60
The Wolseley Motor-Sleigh	60
Mixed Tractors	61
The Lefebvre Tractor	61
Applications of Chain-Track Traction	62
Chain-Track Haulage Wagons	63
Other Applications of the Chain-Track	63
Parsons Excavators	63
Austin-Drainage Excavators	64
Applications of Trench Excavators	64
Back Fillers	64

Appendix I 66
Historical Note 67

Appendix II 74
Authorities Consulted 75
Table 1 Four-Wheel Drive Vehicles, Engines, Speeds Brakes etc. 76
Table 2 Chain-Track Tractors 78

Discussion 84

Communications 93
Table 3 95

Appendix III 105
Bates Steel Mule 106
Table 4 Wheeled Tractors 108

Wheeled Tractors 110

Tanks and Chain Track Artillery 155
December 1921 156
Part I March 1922 184
Part II April 1922 198

Foreword

By Andrew Hills

The name of Lucien Legros may be somewhat unfamiliar to the modern reader, as was the case with so many other talented engineers of the 19th and early 20th centuries, his work did not make his name widely known outside the engineering community. Consequently, in the more than a century since their first appearance, his contribution to the development of tracked vehicles in First World War has not received either sufficient recognition or understanding.

Lucien Alphonse Legros O.B.E., M.I.Mech.E., M.I.C.E.
1865 – 1933

Lucien Alphonse Legros was a renowned engineer in the latter part of the nineteenth century with a particular interest in transport, an interest arising from his experience in civil engineering which often involved moving large amounts of material. This inevitably led to an interest in all forms of mechanized vehicles, especially vehicles capable of moving off-road and an understanding of the difficulties entailed in heavy vehicles moving on a poor-quality surface.

Today it seems obvious that the solution to moving heavy loads across a surface with a low bearing strength usually lies in the use of tracks, but that is with the benefit of hindsight. When, during the First World War, the need for vehicles capable of moving over shell torn ground became apparent, the only mechanically propelled vehicles with which most people were familiar used wheels. Moreover, the use of a wheeled vehicle tended to be confined to a fairly smooth surface, as on other surfaces they fared badly. In these circumstances the 'load' to be carried across this ground would be the weight of men, arms, and armour on the vehicle to protect against enemy fire. This created a hitherto rarely considered problem, as prior to 1915 there were very few heavily armoured vehicles for use on or off roads and those which did exist saw little actual use and achieved no notable success.

It was here, in the spring of 1915, with the realities of a modern war causing enormous casualties, that it was realized that a mechanical solution to the domination of trench-based warfare might be found. Early ideas for solving this problem were going to need to harness the abilities of men like Legros, as it was readily apparent that the issues facing the designers of a vehicle capable of moving cross-county were different to those of a road bound vehicle. The main problems were to prevent the vehicle sinking into soft ground and enabling it to surmount obstacles, problems that were only exacerbated by the need to carry a suitable amount of armour to protect the machine and its crew. In terms of a solution to moving across the ground, the choice was either to use much bigger and wider wheels, or to use tracks and both would be worked on to find the best solution.

It was the experience of engineers like Legros that eventually led the authorities to the conclusion that the answer lay in a vehicle that used tracks rather than wheels. It is, therefore, unsurprising that he worked on the Landships Committee with another pioneer of cross-country transport, Colonel Rookes Crompton, between them bringing to the work of the Committee a wealth of knowledge and experience.

However, while the use of tracks might seem the obvious solution today, the move away from the use of wheels was a difficult concept for many to grasp and as ever, there was resistance to a new concept. Thus, while more far-sighted engineers like Legros were convinced that a tracked vehicle was the way forward, they had to overcome not only the significant technical problems associated with any new machine, but also the prejudice of those in a position to nurture or decry such a new idea. Even when a new idea is welcomed, there are often a plethora of problems to be overcome before the idea becomes a viable reality, all the myriad of details that a complex machine involves have to be worked out and numerous technical problems overcome before success can be achieved.

The coupled Bullock Creeping-Grip tractors during trials. Legros' friend and colleague Colonel Crompton is standing on the far right with the cane.

The tracked vehicles available at the start of the First World War were essentially those which had been developed for agricultural use. They tended to be small as they were basically a mechanical replacement for a horse. Nevertheless, this tracked horse-replacement could, with sufficient far-sightedness, be seen to offer the solution which was sought, it just needed to be made much bigger and be adapted to a very different purpose. As a tracked vehicle got bigger and was re-purposed, new and unexpected challenges would inevitably be encountered and solutions had to be developed to overcome them. The tracked vehicle offered the potential of more power and the ability to move over soft ground. All that was needed was to create such a vehicle and moreover, to create it as quickly as possible.

The Landships Committee was faced, essentially, with a fundamental decision. Should they start from scratch and try to design and then build a vehicle which was, from the outset, matched to their requirements, or should they start with whatever vehicles already existed and try to adapt or develop them to meet their requirements.

The former idea is often the more attractive as it offers the engineers and designers the freedom to incorporate their own ideas. Unfortunately, it also creates a situation where expensive lessons have to be re-learned. The Landships Committee wisely decided to capitalize on what already existed and sought what was available. They determined that the Bullock Creeping Grip Tractor had potential, although it was too small as it stood to meet their needs. Nevertheless, two were ordered from America and connected back-to-back as an articulated vehicle.

Legros and Crompton learned much from this early experimental work. They were able to identify which elements of the original machines could be developed and probably even more importantly, which elements were unsuitable. This knowledge could then be incorporated into their own designs. The key contribution of Legros and Crompton was the realization that in order to be effective on soft or uneven ground and particularly in the surmounting of obstacles, the tracks had to be much longer. It was a simple but absolutely fundamental concept and when 'Little Willie', the first tank, made its debut in the autumn of 1915, it incorporated the longer tracks in the form that Legros and Crompton had selected as being necessary to create an effective tracked machine.

It was an incredible achievement. Starting from an ill-defined notion of what was needed in the spring of 1915 to a vehicle that, while far from perfect clearly pointed to the viability of the concept and was operational in the autumn of the same year, was a remarkable achievement. A developmental process that would normally have taken years had been compressed into a few months.

While the final form of British tank was undoubtedly very different to that envisaged by Legros and Crompton, it nevertheless still relied on using tracks rather than wheels. It is perhaps ironic that the track design which eventually saw combat on those first tanks was a new design purpose built for the task. If that had been too difficult to envisage in the spring of 1915 when work began, then the work of Legros and Crompton helped to make that possible by the autumn of that year. Thus, it can be stated quite reasonably that when Wilson and Tritton were working on their new track system to replace the lengthened Bullock system chosen by Legros and Crompton, that they did so basing the changes to that design on the lessons which had been learned. In this way it can be seen that as in any complex engineering development the final product combines the ideas, the work and the skills of a great many engineers.

It is the bringing together of all these contributions that creates the final product. The tank was just such an achievement. When Sir William Tritton and William Foster & Co. Ltd., of Lincoln demonstrated the first production model of what was to become a crucial weapon in securing the Allied victory, they were presenting the work of not just a few, but many engineers who had all contributed to the achievement. Unfortunately, the critical contributions of these engineers, both big and small has too often been allowed to fade away into obscurity.

The tank, which contributed so much to the eventually Allied victory, was not the invention of any one man, as the evidence placed before the Royal Commission tasked with awarding credit and financial rewards for war-winning developments demonstrated only too clearly. There were numerous claims and counter claims with varying degrees of validity and it is here that a fundamental truth emerges. The tank, like almost any other major engineering project, was the product of many fertile minds; numerous experiments some of which were successful whereas other failed, but all of which contributed to the fund of knowledge on which the final and successful design was created.

Lucien Legros went on to become the President of the Institution of Automobile Engineer from 1916 to 1917, a period when the vehicle powered by the internal combustion engine underwent some of the first stages of a radical transformation. The extensive use made of lorries to serve the needs of the armies, with the cars and buses to transport the fighting men and their supporting services had changed forever the perception of the mechanically propelled vehicle. It was more reliable, more robust and available in far greater numbers thanks to improved production methods.

In 1918, Legros undertook the preparation of a paper to be presented to the Institution of Mechanical Engineers on the topic of traction on poor quality surfaces demonstrated his background knowledge which went into his clearly making reference to his background work in developing tanks. Colonel Crompton even alluded to this in the discussion of the paper as work of *"a very important national matter connected with the subject, of which they were not now allowed to speak"*. The paper is therefore a snapshot in time of development work which was still in progress on the subject.

It would not be until 1921 that Legros was finally able to combine his knowledge of, and role in, the development of the tank and tracked vehicles to produce a series of articles that were published in 'The Automobile Engineer' under the title 'Tanks and Chain Track Artillery.'

However, by 1922, the war had been won and people were anxious to put the horrors of the conflict behind them with the result that there was no longer the same high level of popular interest in military vehicles. The work into which Legros, Crompton, and many others had invested so much energy and talent was no longer in the public eye.

This book is not intended to be a biography of Legros as the material for such a volume is not readily available. Rather it is an attempt to portray his contribution to vehicle development by allowing his own work to reach a wider audience. It is to be hoped, that by understanding the contribution made by these men, whose work has received inadequate attention over the last century, that a more thorough understanding of the way in which these weapons were first unleashed may be achieved.

Publisher's Note

In the first part of publication reprinting Legros' work on the technology of traction on and off road the original has been faithfully reproduced as well as possible. The original manuscript from which it has been taken is over a hundred years old and has suffered from yellowing and warping of pages, and a broken binding. Where necessary some of the photographs and images has been digitally cleaned or manipulated to try and correct as far as possible the yellowing and warping. The text is reproduced as it was originally published with the original punctuation and grammar. Where page numbers within it are reproduced these have been changed to fit the new page numbers in this publication.

TRACTION ON BAD ROADS OR LAND.

BY

L. A. LEGROS, M.I.Mech.E.

(With combined Discussion on Mr. Arthur Amos' Paper
on "Utility of Motor-Tractors for Tillage Purposes.")

EXCERPT MINUTES OF PROCEEDINGS
OF THE MEETINGS

OF

THE INSTITUTION OF MECHANICAL ENGINEERS,
IN LONDON, 18TH JAN. AND 15TH FEB., 1918.

MICHAEL LONGRIDGE, PRESIDENT.

BY AUTHORITY OF THE COUNCIL,
and passed by the Censor.

PUBLISHED BY THE INSTITUTION,
STOREY'S GATE, ST. JAMES'S PARK, LONDON, S.W. 1.

The right of Publication and of Translation is reserved.

TRACTION ON BAD ROADS OR LAND

By

L. A. LEGROS, M.I.MECH.E.

(With combined Discussion on Mr. Arthur Amos' Paper

on "Utility of Motor-Tractors for Tillage Purposes")

EXCERPT MINUTES OF PROCEEDINGS
OF THE MEETINGS

OF

THE INSTITUTION OF MECHANICAL ENGINEERS,
IN LONDON, 18TH JAN. AND FEB., 1918.

MICHAEL LONGRIDGE, PRESIDENT.
BY AUTHORITY OF COUNCIL,
and passed by the Censor.

TRACTION ON BAD ROADS OR LAND.

BY L. A. LEGROS, OF ACTON, Member.

The problems involved in the design of vehicles for travelling, or for the haulage of goods, or agricultural implements, or produce over the land or bad roads, depend for their satisfactory solution upon conditions that vary widely. The worse the condition of the road or land to be traversed, the more essential it is that due account should be taken of the nature of the ground, its inequalities, its resistance to load and to those shearing forces to which its surface is subjected by tractive effort.

The methods adopted for overcoming the difficulties peculiar to such traction problems can be divided into two main classes: the one class is that in which the driving power is distributed over more than two driving wheels; the other class is that in which the tractor drives through chain-tracks which reduce the insistent load. The two classes of tractor are, consequently, applicable to different conditions, and while the methods cannot be considered as competitive, the limitations of each can be best appreciated by direct comparison of their respective peculiarities and advantages.

Under the ideal conditions of traction on rails, the driving power transmitted through a single pair of wheels may suffice to haul a train of many coaches or wagons. Under the less favourable conditions of ordinary roads, a single pair of driving wheels may serve to transmit the power necessary for hauling two or more trailers. Traction on the ordinary road by metal-tyred wheels, however, demands in the driving wheels a large diameter and great width of tread, in order that the insistent load may be reduced to reasonable intensity. How sensitive roads are to the loads that they carry may perhaps be best realized by considering a gravel road, on which ordinary devices can, in normal conditions, be used safely; the same devices may, however, cause serious damage when the load-carrying coefficient is even slightly disturbed, as, for example, by a thaw.

Among schemes for road haulage devised with the object of reducing the load per driving wheel may be mentioned the Renard road-train,* consisting of 6-wheeled vehicles. The central wheels of each vehicle were metal-tyred driving wheels, and the end wheels were fitted with rubber tyres and Ackermann steering.† The leading vehicle formed the power unit, and power was transmitted to the central axles of the other vehicles of the train by an articulated shaft. Under State subsidy the system had a fair trial in France, but it has not developed to any notable extent. The attempts at multiple-axle drive have, however, indirectly led to the development of the four-wheel drive, a system which, while less ambitious, has considerable advantages for some conditions of traction over the ordinary two-wheel drive.

---

* Proceedings Inst. of Automobile Engineers, Vol. ii, pages 8, 18, 27, and Vol. v, page 15; also Appendix I, Historical Note, page 136.
† Engineering, Vol. 83, pages 237 and 408 (1907), Appendix I, p. 136.

PART I.

PART I.
FOUR-WHEEL DRIVE.

The four-wheel drive may be combined with either two-wheel steering or four-wheel steering. An example of the first of these classes, schemes for which were proposed as early as 1898, was the touring car built by Spyker of Amsterdam in 1903, and exhibited at the Crystal Palace in 1904. The Author has already referred to this in the discussion on a Paper by M. Brillié read before the Institution.* M. Brillié's Paper treated of an improved motor lorry of the second of these classes, driving and steering through all four wheels.

The four-wheel drive vehicles of to-day have much in common with the ordinary back-axle driven lorry, or truck, in type and arrangement of the engine and its accessories, such as carburettor, radiator and clutch, and also in the arrangements of gear-box, springing and brakes. Those features only will be described, therefore, in which the four-wheel driven vehicle differs from the ordinary type.

FOUR-WHEEL STEERING.

In the ordinary front-steering vehicle, fitted with the Ackermann axle, the design endeavours to meet the condition that in plan the axes of both front wheels shall intersect at a point on the axis of the back wheels, as shown in Fig. 1 (page 5).† In the four-wheel drive car with four-wheel steering the problem assumes a more general form, and the condition to be met is that the axes of all four wheels shall intersect at the same point, Fig. 2. So long as this condition is fulfilled it is not essential that the inclination of the axes to the normal shall be the same for the front and back axles; but in practice the steering is generally symmetrical, Fig. 3, and the virtual centre about which the vehicle turns is situated on the normal centre line. This results in a reduction of the radius of the turning circle approximately to one half of that obtained with the same front axle and with front steering only. The vehicle has therefore much greater handiness in certain respects, but is subject to the disadvantage that if brought near a kerb, and particularly a high kerb, the wheels cannot be locked sufficiently to permit of taking a small turning circle in. leaving. This can be a distinct disadvantage in cities, though it is to a great extent compensated for by the greater ease with which the vehicle can be manoeuvred backwards.

The locking of the rear wheels restricts the space available for framing the body of the vehicle at the height of the chassis to a narrower width than with two-wheel steering. This does not affect the arrangement of a cylindrical tank or hopper-shaped body, but for the ordinary flat platform it is necessary to give greater height than in normal practice. Although the Author has not found an actual example of the non-symmetrical steering, -shown in Fig. 2, it is to be noted that conversion to this arrangement involves nothing more than altering the lengths of two levers, and that the increased width of body which could be carried direct on the chassis might well compensate for the slight increase in the turning radius.

---

* Proceedings, I.Mech.E., 1914, page 589.
† See 'Road Locomotion,' by Dr. H. S. Hele-Shaw, Proceedings, I.Mech.E., 1900, pages 200 *et seq.* and Plate 32.

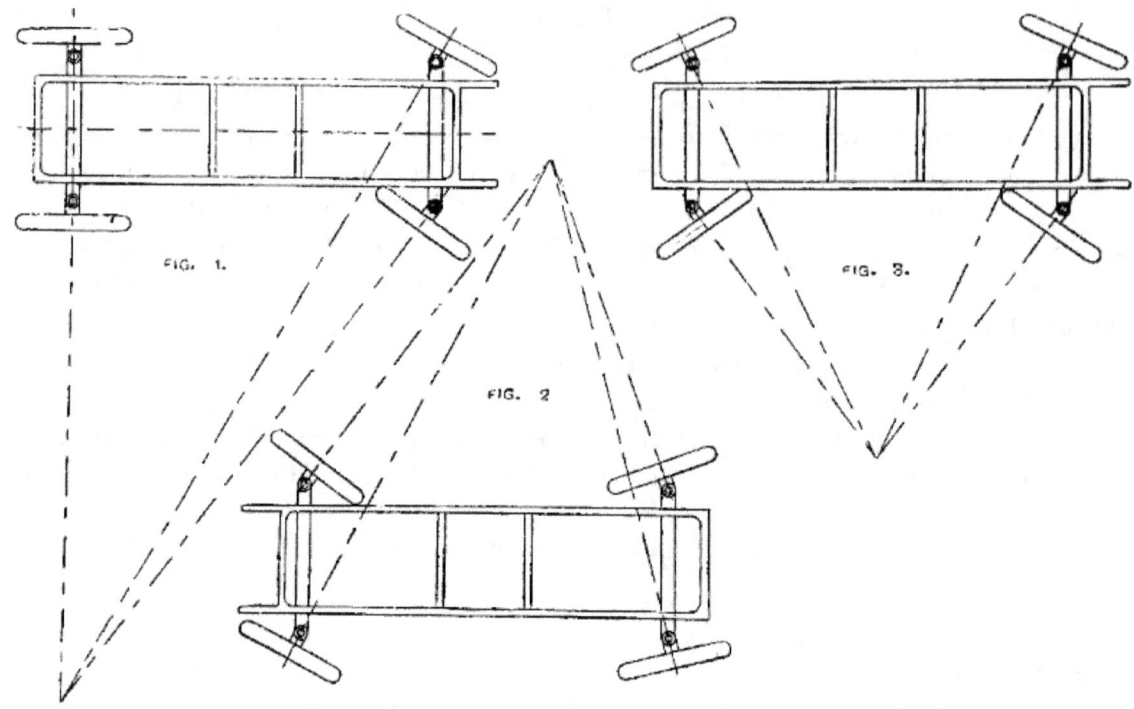

FIG.1. – Ackermann Front Steering.
FIG.2. – Four-wheel Steering; Non-symmetrical.
FIG. 3. – Four-wheel Steering; Symmetrical.

For driving over country roads or fields with many irregularities or obstacles the four-wheel drive has undoubted advantages due to the fact that the percentage of load available for adhesion is nearly double that of a two-wheel drive.

FOUR-WHEEL BRAKING.

The application of brakes on automobile vehicles running on ordinary or bad roads involves inevitably the possibility of side-slip. It is well-known that, when the coefficient of friction for lateral displacement of the tyres is small, and the brakes are applied too hard on a vehicle with rear wheel brakes, side-slip will result, and the back of the vehicle will 'come round' unless prevented by the skill of the driver. The extent of the slip can be reduced by releasing the brake and steering to the side towards which the back wheels are moving. When ordinary side-slip takes place on a cambered road, and the back of the vehicle is tending to approach the kerb, the difficulty of arresting the movement is increased for two reasons; first, because the front wheels must be steered towards the kerb, thus reducing the space available for correction of the error of movement, and secondly, because the camber of the road tends to increase the lateral velocity of the vehicle. The paths taken by the wheels respectively in an ordinary rear-wheel side-slip, on a surface which is level transversely to the direction of movement, are approximately as shown in Fig. 4, in which for the sake of clearness the front wheels are shown narrower in track than the back wheels.

FIG. 4. - *Path of Wheels in 180° Side-slip.*

There is, however, another form of side-slip which is much more dreaded by the expert driver, as it is harder to check; this is front-wheel slip.* On greasy asphalt, wet oolite, and damp clayey roads, under normal running conditions the driver may be warned in time by the failure of the vehicle to respond to movement of the steering wheel. With this form of slip, application of the brake, resulting in the transference of some of the load from the back to the front wheels, may produce a sufficient change in the conditions to enable the path of the vehicle to be corrected. It is necessary to consider this aspect of the problem in motor vehicles that are rear-driven. As usually designed the axle loads are approximately equal when the vehicle is stationary; but this condition is changed when the vehicle is running, for the driving torque, corresponding to the speed of the periphery of the road wheels, acts on the back axle and relieves the front axle of part of the static load. Practical recognition of this fact is seen in the racing car, the evolution of which has shown a progressive increase in wheel-base accompanied by forward displacement of the centre of gravity of the car. A racing car of 10 foot wheel-base with a 70 b.h.p. engine, when running at 80 miles per hour, will have a torque at the back axle of about 353 ft.-lb., but the same vehicle if climbing a hill with a gear ratio of 3.5 to 1 (reducing the speed to 23 miles per hour) will have a torque at the back axle of 1,235 ft.-lb., transferring about 1 cwt. from front to back wheel. If, on the other hand, the rear-wheel brake is applied on an ordinary vehicle, the reverse torque so set up may easily be much larger than this; and the resulting change of load conditions on the wheels may therefore increase the insistent load on the front wheels sufficiently to obtain the necessary friction between road and wheel for steering. It is obvious that front-wheel driving implies heavier loading on the front wheels, and involves greater difficulty for the driver in correcting front-wheel side-slip should it occur.

For stoppage under conditions of good road grip the four-wheel brake acting equally on all four wheels is ideal; by the use of compensating gear the braking action can be distributed equally over the four wheels, and for any given speed the car can be stopped in a much shorter distance than with two-wheel braking. When, on the other hand, the surface is greasy, a skilled driver has greater control over a vehicle braked only on the back wheels.

* Proceedings, Inst. of Automobile Engineers, Vol. ii, pages 57-163.

FOUR-WHEEL DRIVING.

Starley's application of the differential gear to the back axles of vehicles as a means of compensating for inequalities in diameter or in travel of the wheels requires further extension in the case of the four-wheel drive. When the vehicle is driven mechanically by a single petrol motor, it is not only necessary to provide a differential gear on each axle, but, when these are of the ordinary bevel or spur types, a third differential gear must also be interposed between the two axles to compensate for differences in the mean diameters, and consequently in the revolutions made by the respective pairs of wheels. This introduces a further complication in the problem of driving. In the ordinary two-wheel drive if one driving wheel is held and the other is off the ground, the latter will revolve, other things being equal, at double the speed at which the pair would revolve if both were free; whereas, if with the arrangement of three differentials three wheels are held and one wheel is off the ground, this wheel will revolve at four times the mean speed at which the four would revolve if free or driving normally. Conditions under which the load on any one wheel is so far reduced as to allow of slipping occur frequently on very uneven ground owing to release of spring pressure, and they also occur on ground presenting considerable fluctuations in coefficient of friction. To enable the four-wheel drive vehicle, fitted with the usual bevel or spur wheel types of differentials, to work under these conditions it has been found necessary to fit a differential-lock to the main or central differential on the longitudinal driving shaft connecting the two axle drives.

This difficulty is not so serious in vehicles fitted with electric drives, in which it is, of course, possible to fit, a separate motor to each wheel. An arrangement of this kind is used in the Couple-gear car, and is shown in Fig. 32, Plate 6.

AUTOMATIC DIFFERENTIAL LOCKING GEARS.

The problem of power transmission when one wheel is off the ground, or has no hold, as when in a clay-hole or on ice may, however, be solved in another way. The slipping or free-running of one wheel is due to the fact that the ordinary bevel or spur differential gear is perfectly reversible. If one wheel is jacked off the ground it will revolve at twice the normal speed corresponding to the engine-revolutions; and vice versa, if the wheel is turned round by hand the engine will make half the corresponding number of revolutions. An attempt made by Hedgeland, about 1905, to solve this problem proved partially successful;* but a really practical solution appears to have been found in the worm-gear differential

The M. & S. differential, shown in Fig. 5, is of this type.† In this arrangement each axle-end carries a crown worm-wheel of the same hand, each gearing into a worm-pinion carried in the differential-box; the two worm-pinions are geared into an intermediate worm also carried in the differential-box. In practice the arrangement of secondary and intermediate worms is duplicated as shown in the figure.

The inclination of the teeth is selected so that the worm-pinion can be driven by the crown worm-wheel, but that the action is not reversible.

Hence the gear will admit of the road wheels travelling at different speeds if both are in contact with the ground, but if one wheel is off the ground or has no hold, the irreversibility of the drive causes the whole arrangement to behave as though it were a solid axle.

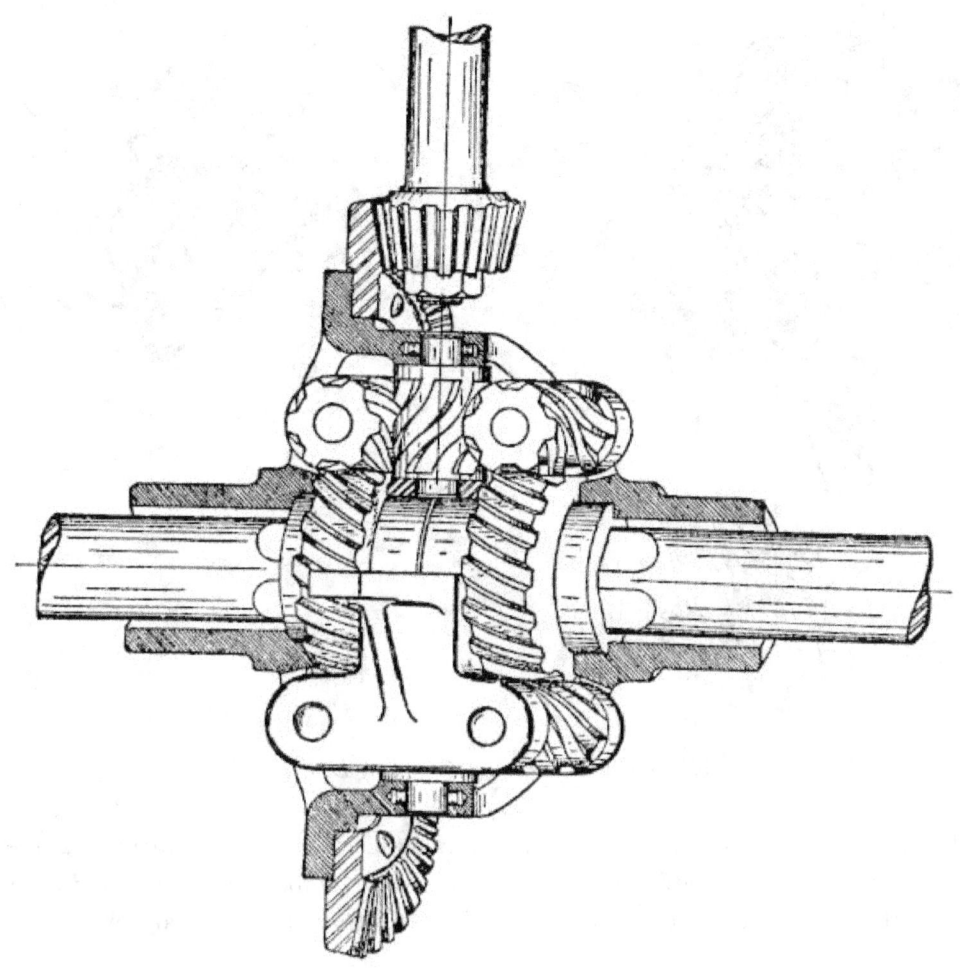

FIG. 5. - *M. & S. Differential.*

---

* *See* Historical Note, page 67.
† *See* the Paper on Differential Substitutes by D. D. Ormsby. Transactions of the (American) Society of Automobile Engineers, Vol. xi, 1916, Part II, pages 288 - 299.

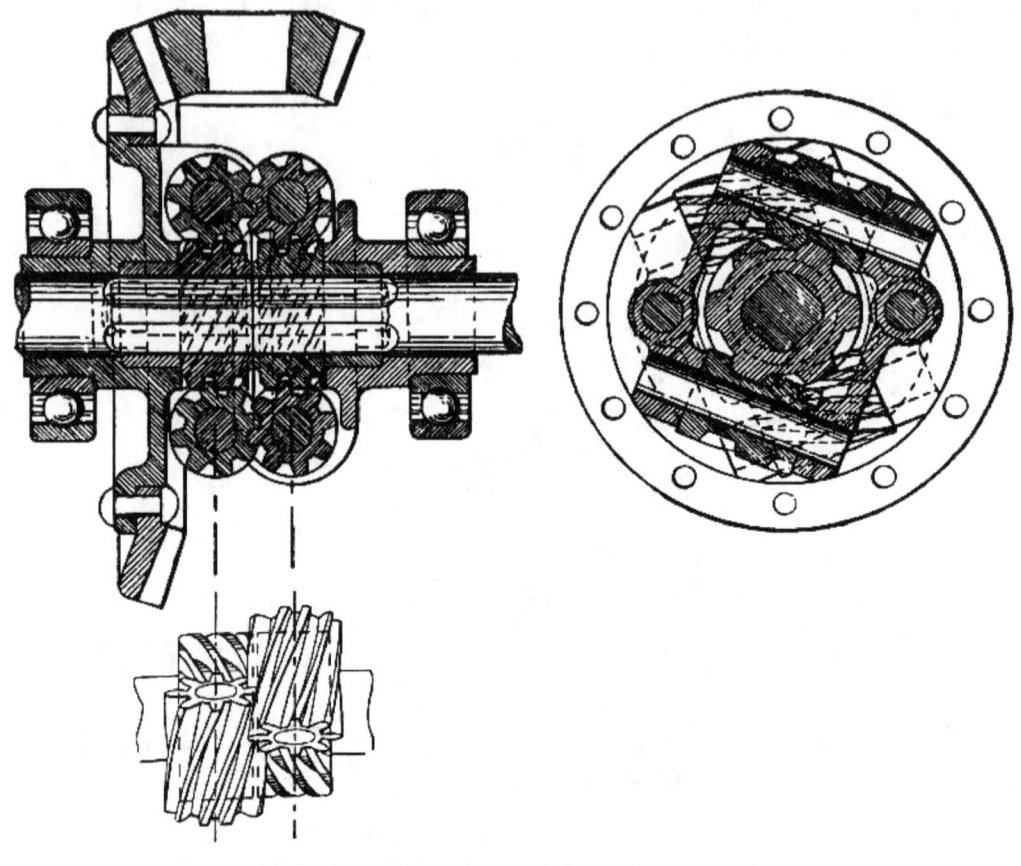

FIG. 6. - *Walter Automatic-locking Differential.*

This peculiarity, causing the worm form of differential to be known sometimes as the "positive – drive" or "automatic locking differential," renders unnecessary the central or third differential, with its differential locks, required when differentials of the bevel or spur types are used for four-wheel drives.

The Walter automatic differential lock, Fig. 6 and Fig. 25, Plate 5, is also a worm-gear differential. In this form the half-axles each carry a crown worm-wheel, and as in the M. & S. gear both are of the same hand. Each crown worm-wheel engages with a worm-pinion carried in the differential-box; these worm-pinions being both of the same hand are carried on axes, the one of which is advanced in relation to the other at an angle double that of the inclination of the worm pinion thread to its axis. This arrangement eliminates the intermediate worm-pinion necessary in the M. & S. gear and thus shortens the overall length.

In practice there are two pairs of worm-pinions as shown in Fig. 6. It is claimed that a static lock is ensured between the two axles" because a worm-gear cannot start driving a worm having a 'much' smaller angle of lead," but that once started the driving is performed at high efficiency," * that is to say, there is differential driving of the wheels, but not equal distribution of power to the two wheels. It is also stated that this design is so compact that it can be substituted for the ordinary bevel or mitre gear differential.

In this country the problem of the automatic differential lock has been attacked by Wolseley Motors in a different way. In one example small multiple-plate clutches are arranged in the differential-box at each side so that they are acted upon by the end-thrust of the gears; this end-thrust exists already in the case of bevel or mitre gear differentials, but in the spur form of differential it is obtained by making the satellite pinions (with axes parallel to the differential axis) alternately right- and left-handed worms geared together and gearing respectively with left- and right-handed worm-wheels, of the same tooth inclination, carried on the respective ends of the axle-halves; the end-thrust of these worm-wheels acts on the multiple-plate clutches in the same manner as in the bevel or mitre gear differentials.

The possibility that the automatic action may not ensure complete locking of the wheel or wheels is provided for in a modification of the arrangement. In this form the multiple-plate clutch sleeves carrying the thrusting wheels (whether bevel or worm) are provided with collars against which forked levers pivoted to the differential axle casing can be made to press by means of a hand lever acting through appropriate linkwork. The gear thus forms a partial lock under normal conditions and becomes a complete lock when desired.

Another form of differential locking automatically is the Dorr Miller, Fig. 7, Plate 1.** This gear consists of the following parts: a central driving plate, eight steel balls carried in holes in the driving plate, two groove-blocks with axially waved races for the balls, and two friction plates, one on the back of each groove-block, having one face bearing on the groove-block, and the other on the inside of the differential-box. As soon as spinning or skidding starts, the groove-block which tends to move faster also tends to move away from the central driving plates by the action of the balls in the waved grooves.

This action causes the outer surfaces of the groove-blocks to bind against the inside surfaces of the friction-plates which in turn tend to bind on the inside of the differential-box, thereby converting the differential temporarily into a solid unit. Under normal conditions it is claimed that the rolling of the balls over the waved grooves permits perfect differential action.

It is the Author's opinion that it is desirable that comparative tests of the various forms of automatic differential-lock should be made by an independent authority, such as the National Physical Laboratory.

---

* The word much has been inserted by the Author in the description here given, because the context implied it. The conditions for irreversibility are given by Lanchester in his Paper on Worm Gear. Proceedings, Inst. of Automobile Engineers, Vol. ii, pages 219 *et seq.*
** The Author is indebted to Mr. A. A. Remington, M.I.Mech.E., Chief Engineer of the Wolseley Co., for the description of this gear and of the two forms of Wolseley gear.

FOUR-WHEEL DRIVE, FRONT STEERING.

The F.W. D. (Four-Wheel Drive) car of Clintonville, Wis., Fig. 12, Plate 2, is a four-wheel driven car with front steering, and is fitted with hand-operated differential-locks to the central differential, Fig. 10, Plate 1. A skeleton plan of the arrangement of engine and drive is given in Fig. 8, and details of the front-wheel driving and steering arrangements as seen from the front of the vehicle are shown in Fig. 9 (page 12). The driving gear is shown in phantom view in Fig. 11, Plate 1. A dog-clutch is provided for locking the central differential, and making it inoperative as a balance-gear between the two axles.

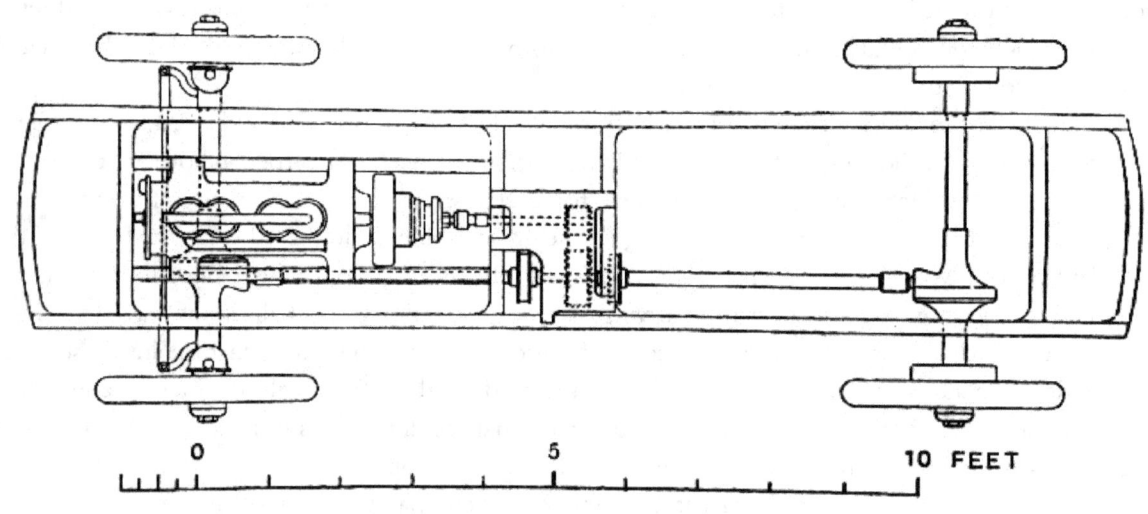

FIG. 8 – *F.W.D. Chassis; Skeleton Plan.*

With the central differential locked, one or other of the clutches on the short shafts, which transmit power to the front and back wheels respectively, can be disengaged by the corresponding dog-clutch; one of these clutches is shown disengaged on the right of the chain-driven differential in Fig. 11, Plate 1. By these means the vehicle can be converted temporarily into a two-wheel back-drive or front-drive vehicle, as may be best suited to the difficulty requiring the use of the differential-lock. Actually two locking gears are provided, one for each end of the differential, Fig. 10, Plate 1. For continuous work as a two-wheel drive car the lock is engaged which is next to the shaft that is being used in driving, so as to avoid transmitting the locking action through the teeth of the differential gear. The position in which the differentials are placed out of the centre of the car, Fig. 8 (page 11), is in part constructional, and in part is selected with the object of obtaining greater central ground clearance than would be possible with a central drive. The back-axle drive of this car is of the kind known as the "full floating type,"* that is to say, the axle is one in which the wheel-driving, or differential shafts, transmit only torsional effort, stresses either by the weight, of the vehicle or by the tractive resistance.

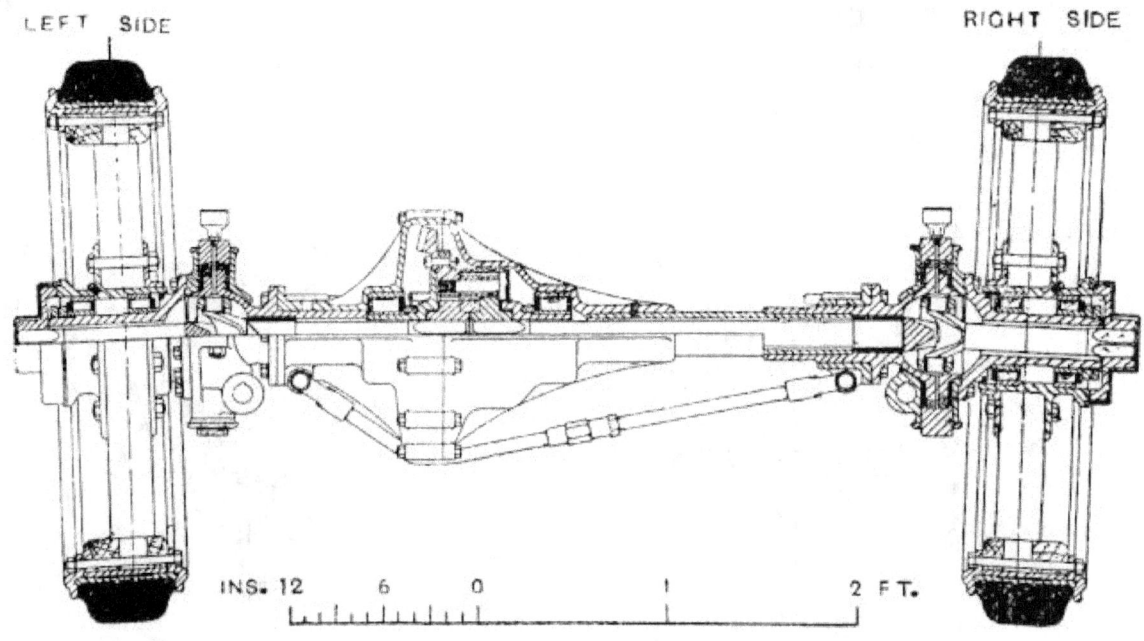

FIG. 9. - *F. W.D. Chassis; Front Axle.*

The front-axle drive is shown in Fig. 14, Plate 2, and a complete side-tipping wagon in Fig. 13, Plate 2. Examples of wagons at work are shown in Figs. 15 to 17, Plate 3, and an end-tip wagon in Fig. 18, Plate 4.

* Live Axles for Commercial Vehicles, by G. W. Watson. Proceedings, of Automobile Engineers, Vol. x, pages 165 *et seq.*

FOUR-WHEEL DRIVE, FOUR-WHEEL STEERING.

The Super-quad of the Walter Motor Truck Co. of New York, Figs. 23 and 24, Plate 4, is an example of four-wheel drive and four-wheel steering, and is shown in plan and elevation in Fig. 19 (page 13); a semi-tractor is shown in plan and elevation in Fig. 20.

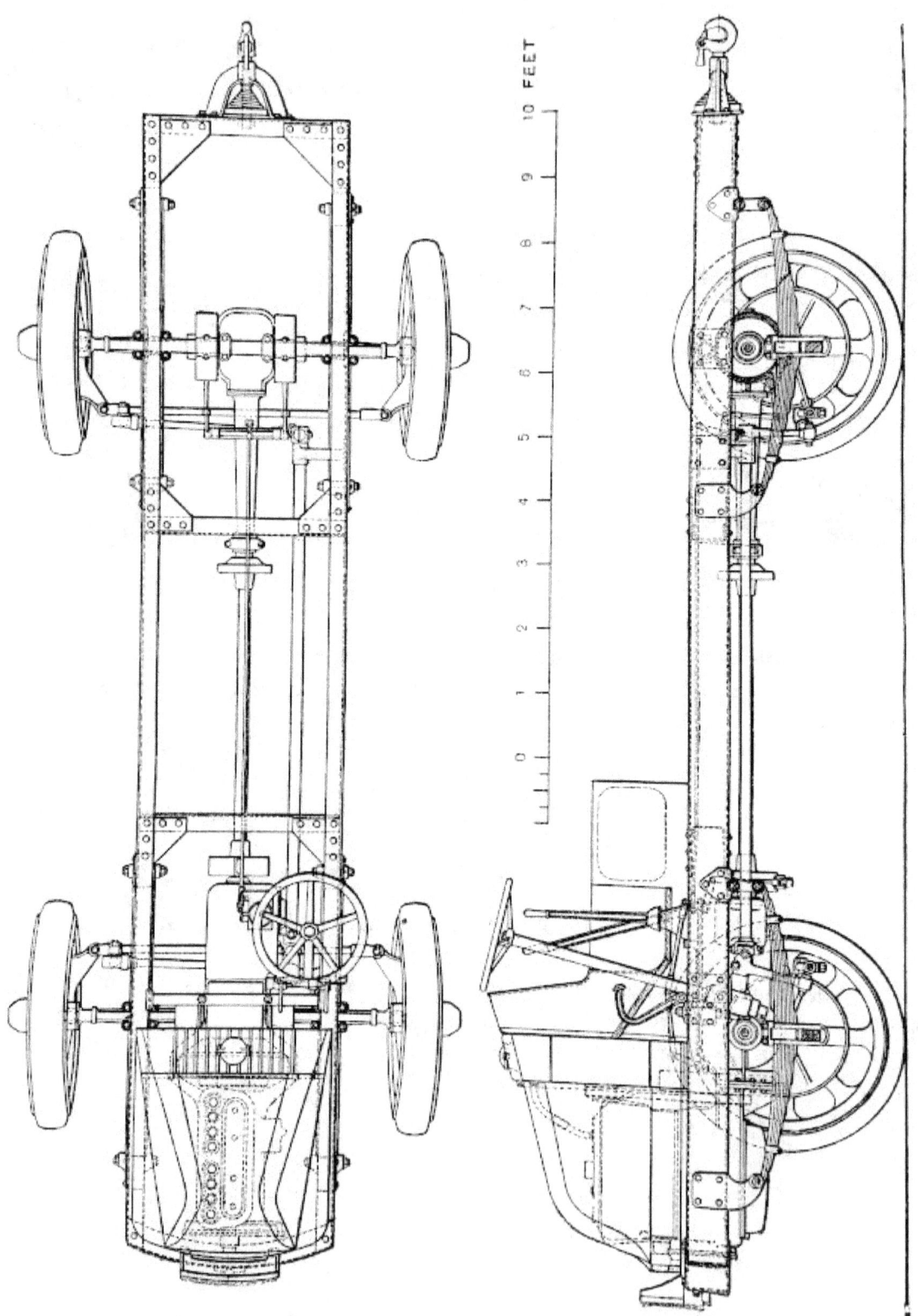

FIG. 19 – *Walter Super-quad; Plan and Elevation.*

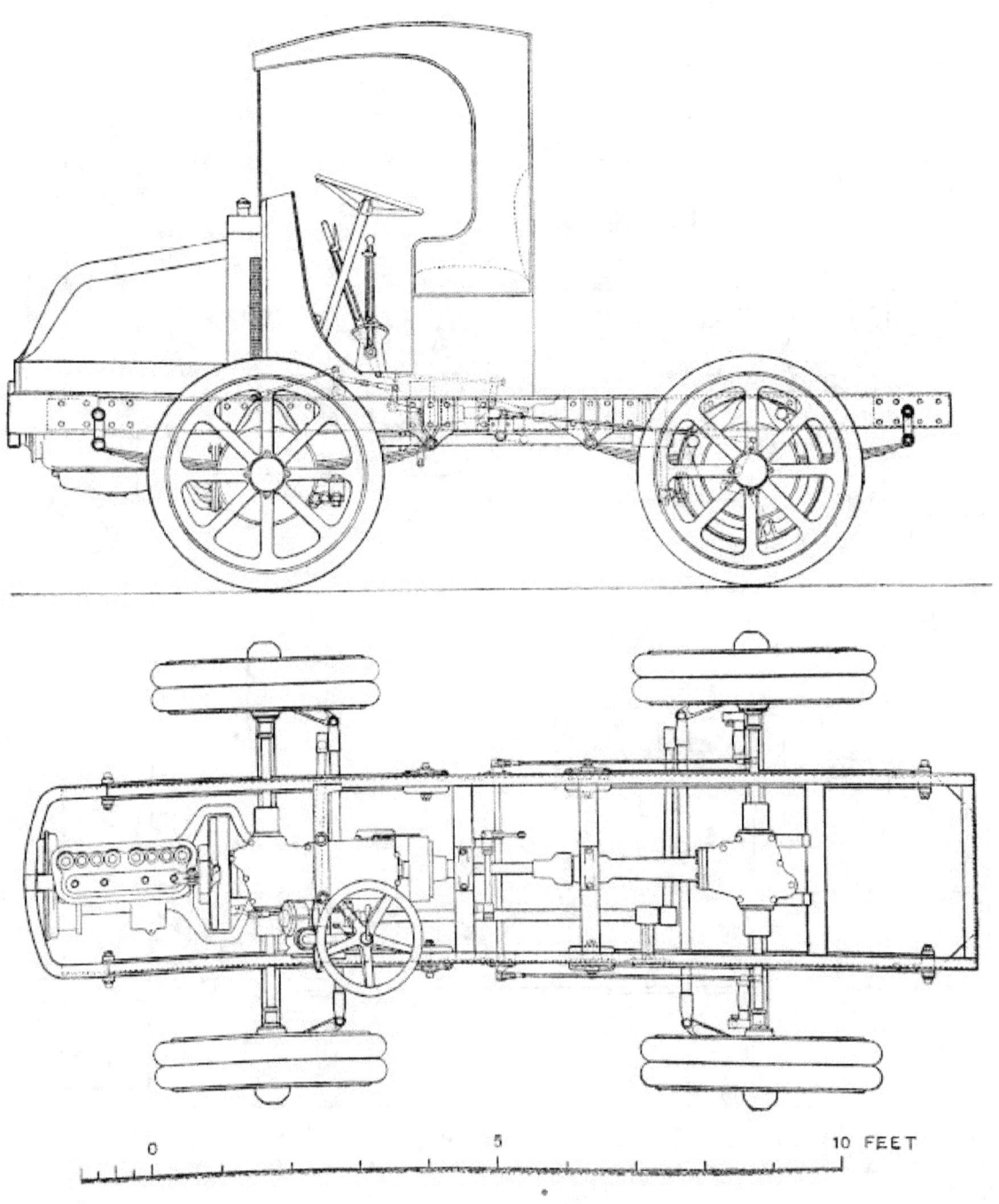

FIG. 20 – *Walter Super-quad; Tractor.*

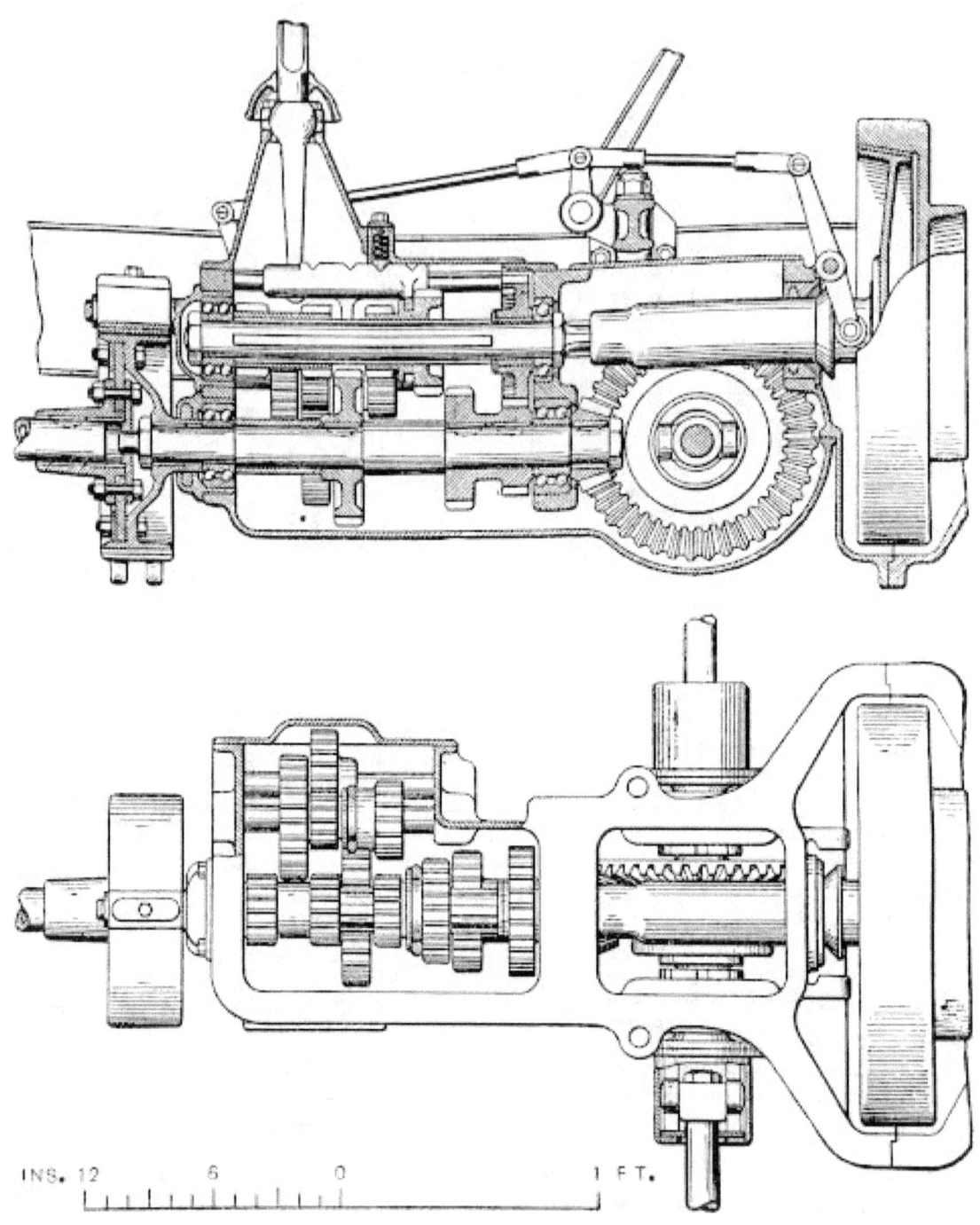

FIG. 21 – *Walter Super-quad. Transmission.*

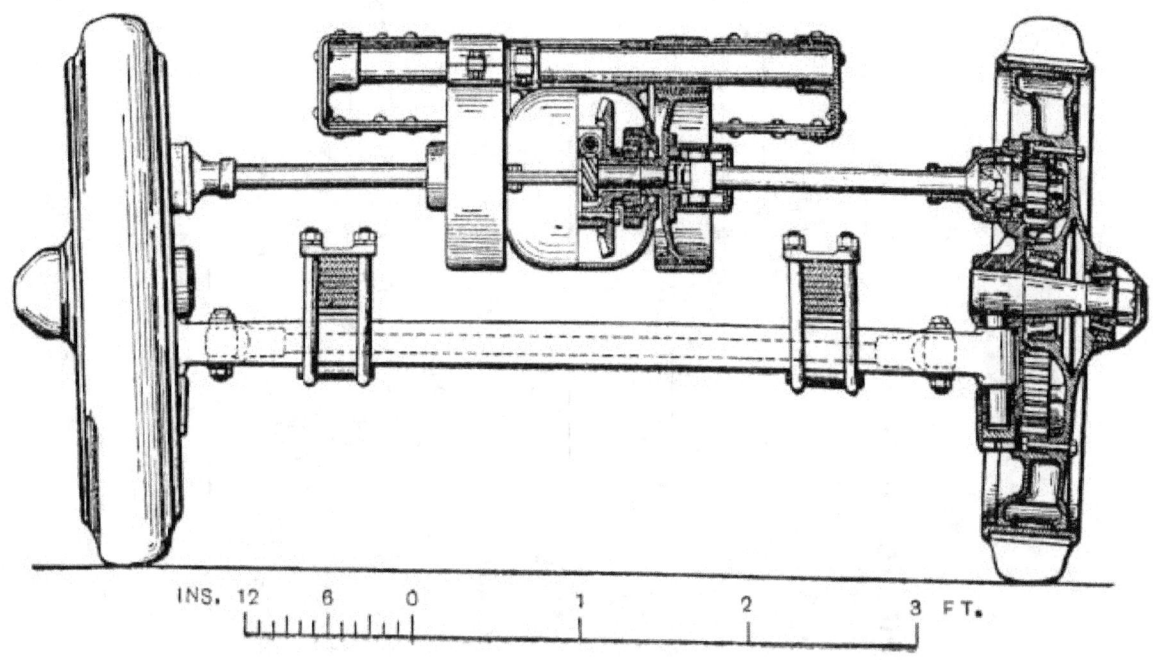

FIG. 22 - *Transmission and Wheel-drive.*

The transmission gear is shown in elevation and plan in Fig. 21, the front-wheel differential shafts with the cardan joints for the short driving shafts to the front wheels being shown in section and outside views. The method of driving the wheels by short articulated shafts is shown in Fig. 22.

Each axle is fitted with an automatic locking differential of the Walter type, which remains automatically locked for normal running, but becomes unlocked when necessary, for example, in turning the car. The two Walter automatic-locking differential gears, Fig. 25, Plate 5, are, moreover, sufficient; the third differential gear on the main shaft, necessary in the case of ordinary differentials, is not required with this device, as either or both wheels on the axle which tends to overrun the other will become unlocked automatically, and will lock again as soon as the distribution of velocity becomes normal.

The vehicle is fitted with a pedal-operated transmission brake and a hand-lever operated back-wheel brake, both of which, owing to the fact that the main differential is either in gear or is locked, also act on all four wheels.

THE JEFFERY QUAD shown in Figs. 26 and 27, Plate 5, is a four-wheel driving and four-wheel steering vehicle, better known in this country than the Walter. This vehicle is fitted with the M. & S. locking differential on each axle, Fig. 5 (page 8).

Some of the forms of body adapted to this chassis are shown in Fig. 28 (page 17), and comprise two forms of hoist or tip bodies and two forms of special bodies for the transport of liquids, etc.

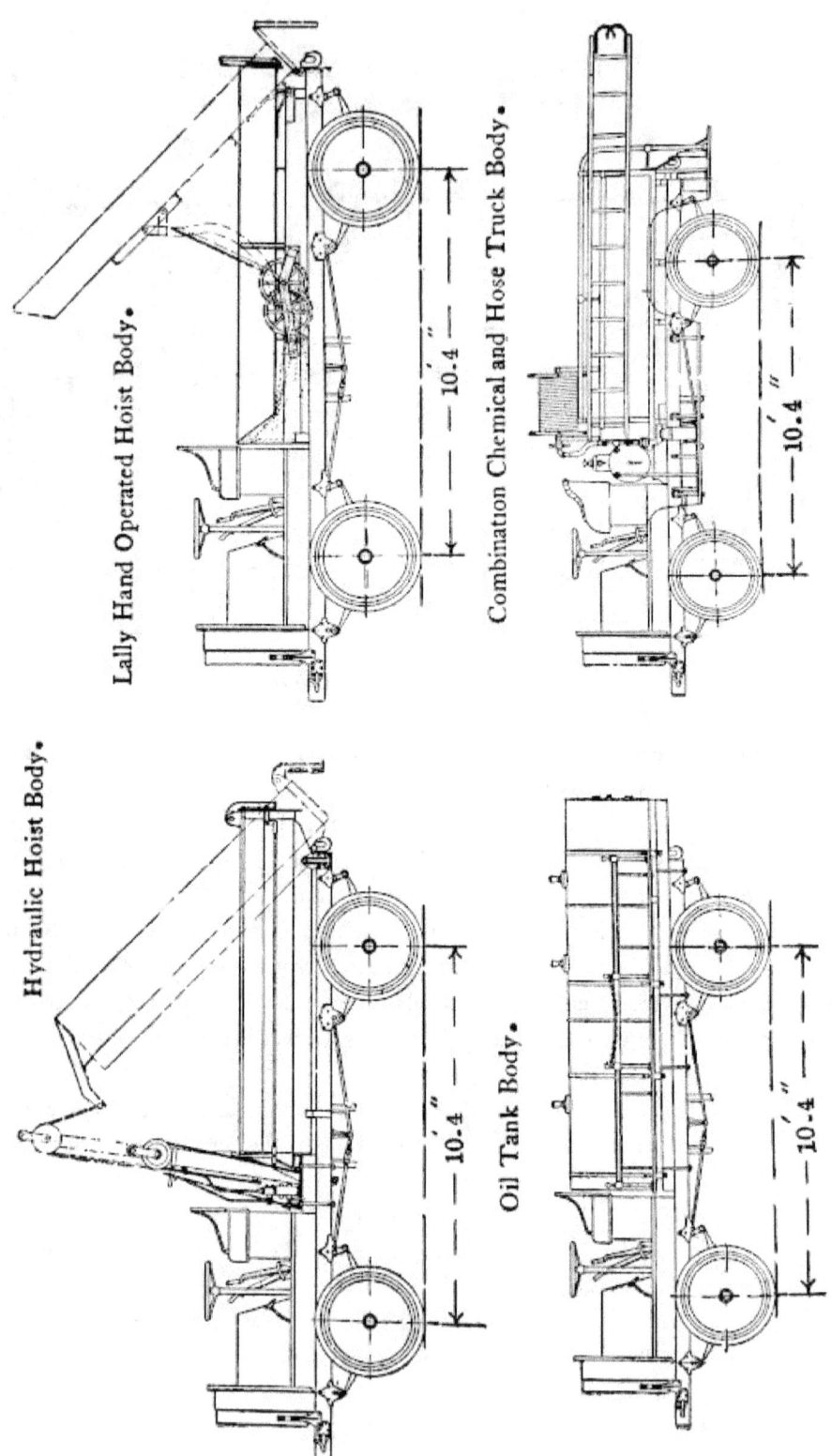

FIG.28. – *Jeffrey Quad; Applications.*

FOUR-WHEEL INDEPENDENT DRIVE WITH FOUR-WHEEL-STEERING.

The Couple-gear electric truck, Fig. 29 (page 18), and also the petrol-electric truck, Fig. 30 (page 19), of the Couple-gear Freight Wheel Co., of Grand Rapids,- Mich., afford examples of four-wheel independent drive with four-wheel steering.

The battery-driven car is shown in elevation, front elevation and plan in Fig. 29. The wheel is of peculiar construction, having two large-diameter bevel wheels enclosing the motor, which is placed out of square with the axle to a sufficient extent to allow both ends of the motor shaft to carry a driving bevel pinion. One of these pinions gears with the outer half of the wheel and the other with the inner half, so that the load on the motor shaft is balanced, a result further ensured by the introduction of an 'evener' or compensating gear: the arrangement of the wheel parts is shown in Figs. 31 and carrying twin solid rubber tyres carrying disks forming the wheel. The gear reduction from motor to wheel is 25 to 1; the relation between tractive effort and speed is given in Fig. 33.

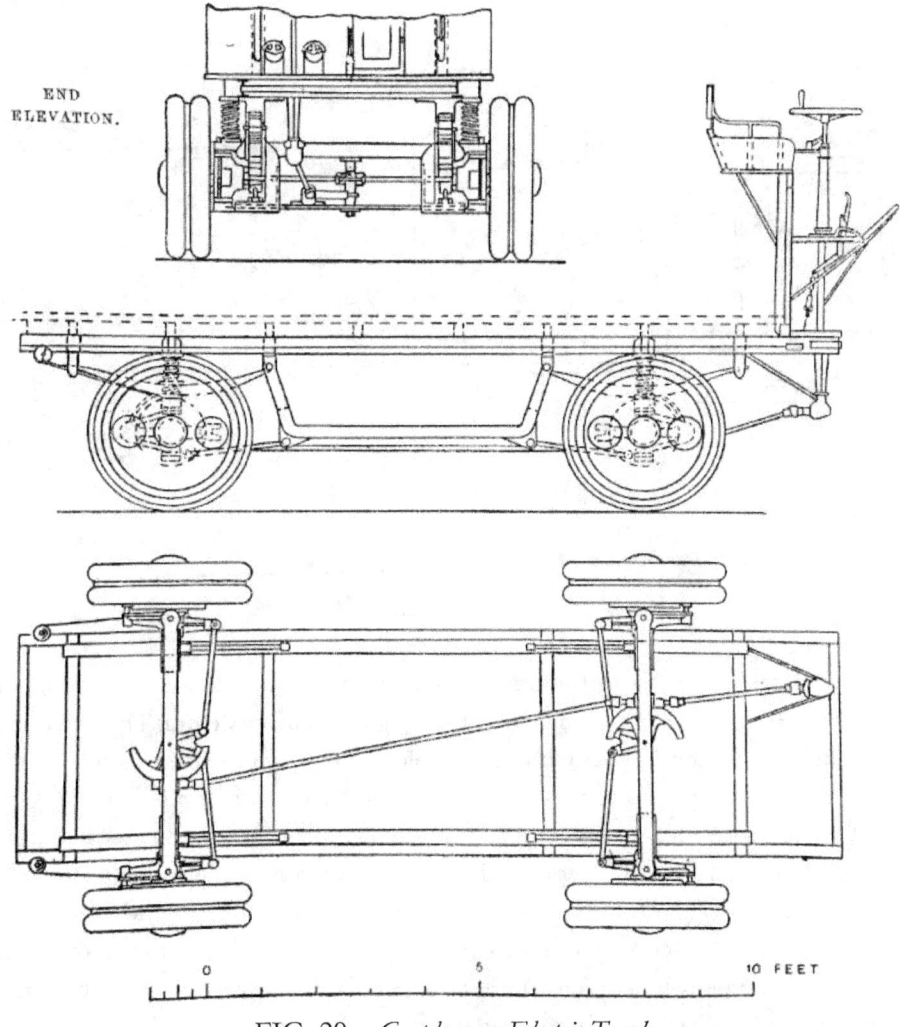

FIG. 29. - *Couple-gear, Electric Truck.*

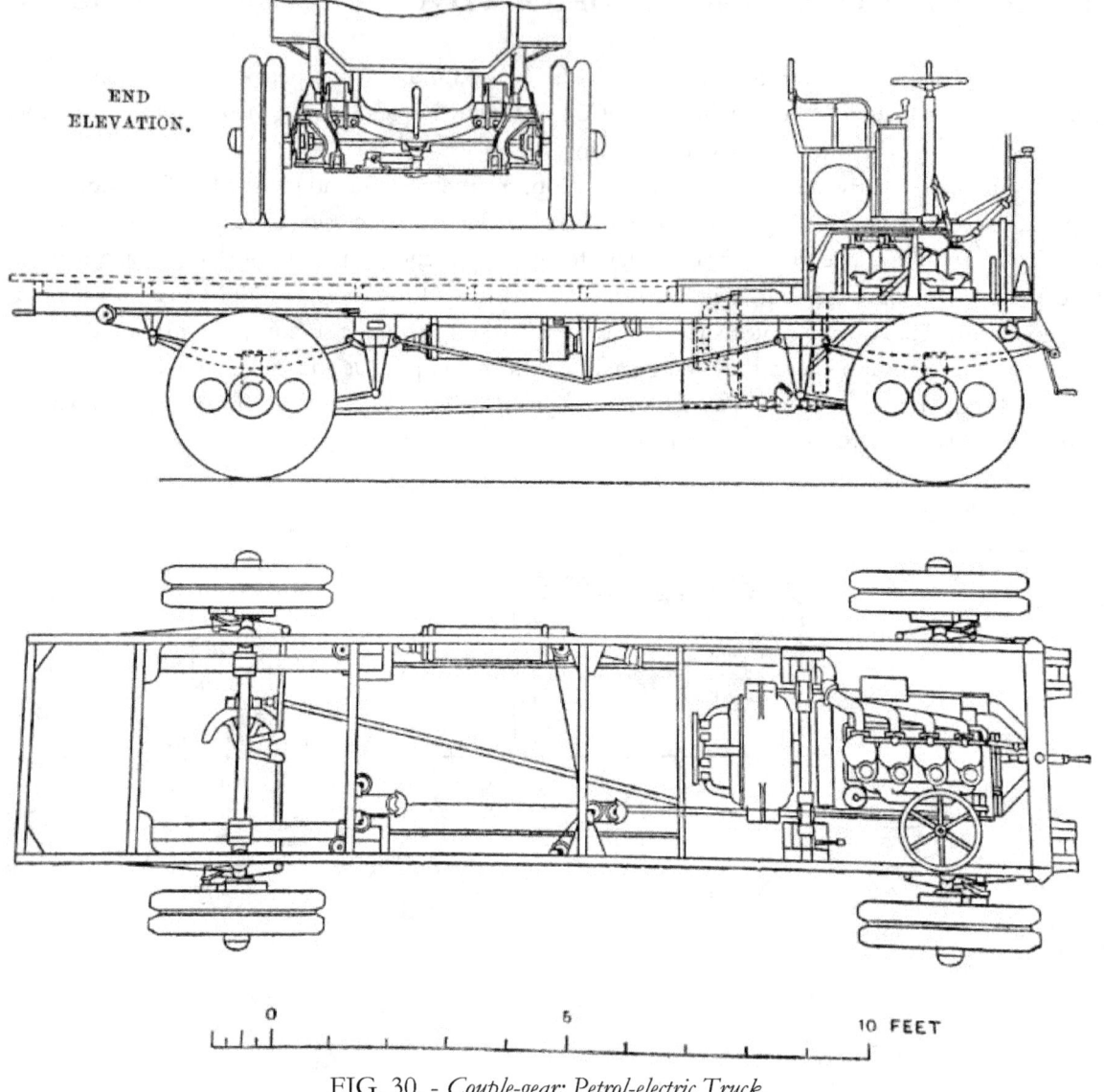

FIG. 30. - *Couple-gear; Petrol-electric Truck.*

The motor is carried on a short hollow stub-axle keyed to a taper seat in the steering knuckle and secured by a nut; the wires are led to the motor through the stub-axle pin. The wheels run on roller bearings and are fitted with band holes with removable doors to give access to the motors. There is one brake of the contracting type to each wheel, and also an electric emergency brake.

To enable the vehicle to be driven over roads which present uneven surface or very unequal resistance to traction a special form of control is arranged. Current is supplied to the motors of the four wheels through a controller of the street-car type modified to suit the conditions.

This controller has five contact points, two of which (the third and fifth in the battery car, and the fourth and fifth in the petrol-electric car) are arranged for what the makers term 'parallel speeds,' the motors being put in parallel when these contacts are made; thus, if one wheel slips it has no effect on the other three, or if two wheels slip there is no effect on the other two. The other three contacts are for what the makers term 'building-up speeds'; on these points a sort of differential action is provided between the two groups of wheels taken diagonally on the car. The right front wheel and left back wheel form one group, and the left front and right back form the other group. When running on the third building-up speed a differential action occurs between the wheels of these groups, so that if one wheel of the group slips, its fellow will be left without power. If under these circumstances the slipping continues, and the driver of the vehicle moves the controller forward to the next contact, all the wheels are then put to work independently of each other.

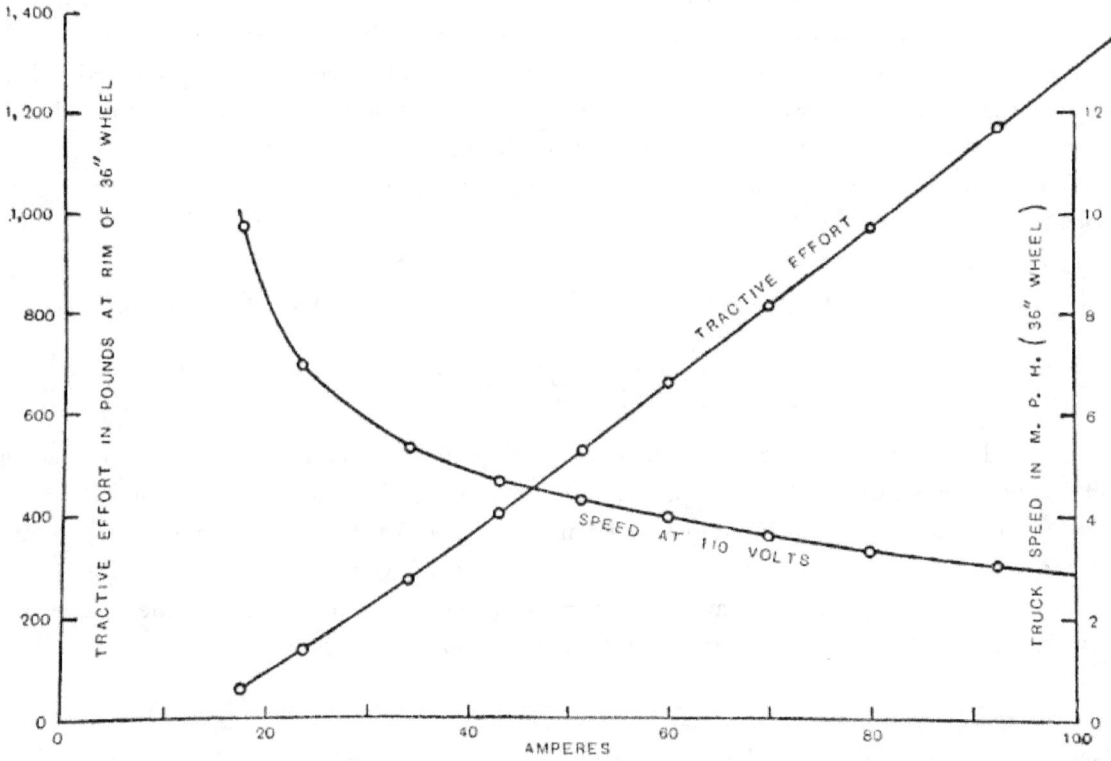

FIG. 33. - *Couple-gear; Diagram of Tractive Effort and Speed of Motor.*

It is stated that on a dry gravel road a tractive effort equal to 60 per cent. of the total weight of the vehicle can be obtained before the wheels begin to slip. This figure depends, of course, on the character of the road surface.

The accumulator-driven type of tractor is only recommended in cases where charging facilities are obtainable, and the total daily mileage does not exceed 40 miles. The petrol-electric vehicle is shown in elevation, front elevation and plan in Fig. 30 (page 19), and the power plant in Fig. 35, Plate 7. These vehicles are not only used as ordinary lorries, but in many instances are adapted with a pivotal carriage to take the front end of a trailer which may be a goods wagon, fire-escape, or tower wagon. In this application the vehicle is sometimes known as a semi-tractor. The Couple-gear driving system is also applied to two-wheel drive vehicles such as fire-engines. Fire-escapes are dealt with by two methods; either as a very long wheel - base vehicle, Fig. 36, Plate 7, or on the semi-tractor principle with a long trailing wagon.

APPLICATIONS OF FOUR-WHEEL DRIVE TO BAD ROADS OR LAND.

In the mining and agricultural districts of North America vehicles are required to negotiate long stretches of sand or mud. They are required to work on desert land, through sagebrush and among sand dunes, under such conditions as are illustrated by the views of F. W. D. vehicles given in Figs. 15 and 16, Plate 3, and they may even be used to haul trains of tip-wagons on a light railway straddled by the tractor, Fig. 17, Plate 3. An end-tip wagon with hydraulic tipping gear is shown in Fig. -18, Plate 4.

The limitations in all these cases are defined by the engine power, transmission efficiency,* insistent load, wheel diameter, tyre width, weight of vehicle, and, when a trailer is hauled, the drawbar pull. There is no satisfactory method at present known for estimating the resistance presented by land, sand or mud, so soft that the wheels can sink to a depth of several inches, though it is known that the resistance under such conditions may be as much as 15 to 20 per cent of the load. The illustrations must therefore be taken as the best method of affording some idea of the work that can be performed by four-wheel driven vehicles.

Comparative figures for the different types of vehicle, their horse-power, speeds, load, weight, wheel-base, track-width, turning radius, and overall dimensions are given in Table 1 (pages 76-77).

CONCLUSIONS.

For the transport of goods over bad roads on gradients, varying from 1 in 15 on roads in which the tyres sink 2 to 3 inches in depth to 1 in 5 on hard roads with bad surface, and for speeds varying from 1.5 mile per hour on grades to 12 miles per hour on fairly level roads, the four-wheel drive tractor has great advantages over the ordinary two-wheel drive tractor.

In the Author's opinion it may be expected to rank as an important factor in the development of districts overseas not far removed from rail-head, but having only primitive roads.

---

* The efficiency of gear-box and bevel transmission does not appear to have been investigated thoroughly. The Author believes that in vehicles of the types described it amounts to about 85 per cent of the engine b.h.p.; Proceedings Inst. of Automobile Engineers, Vol. iii, pages 357 *et seq.*

PART II.

PART II.
CHAIN-TRACK TRACTORS.

The invention of the endless track or self-laying railway dates to nearly a century and a half ago. The first mention of a scheme analogous to that now adopted in so many forms of tractors is the British invention of Richard Lovell Edgeworth (15th February 1770) for a "portable railway" or artificial road to move along with any carriage to which it is applied. Although no drawings can be traced of the invention, the description given is remarkably clear and precise; in fact, almost unaltered, it applies to the greater number of chain-track tractors of to-day. *

The Motor Car Acts of 1896 and 1903, while affecting the restrictions on light locomotives, had little if any direct influence on the development of chain-track tractors. This type of vehicle owes its development to the very great difficulties presented to wheeled tractors by bad surfaces of road or land. The worst conditions are snow, ice, sand, clay, and marsh. Of these snow and marsh have had most often to be faced, the former for the haulage of lumber in winter over snow and ice tracks in forest country, and the latter for the haulage of gang-ploughs through the marshy deltas of California and in the swampy country of Illinois and Wisconsin.

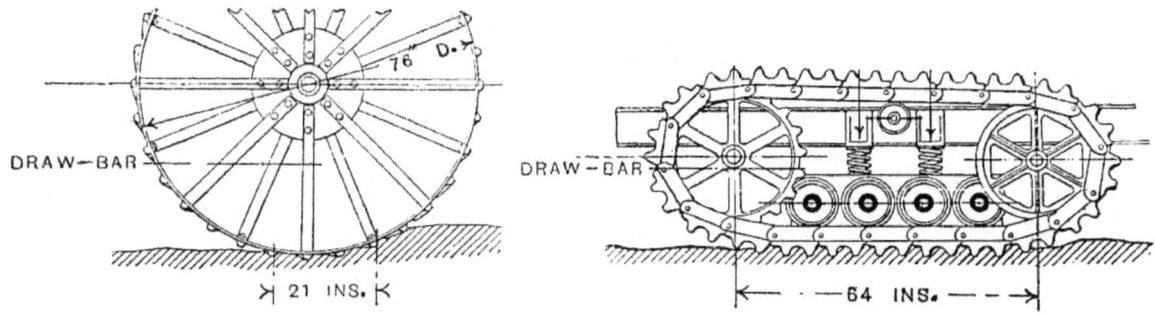

FIG. 37. - *Wheel and Chain-track Compared.*

The necessity for travelling over various kinds of land in moving from place to place has demonstrated the capacity of these vehicles for haulage across alkali and other deserts as well as over clay. To enable chain-track tractors to run on soft or swampy land the insistent weight has been reduced much below that imposed by a horse, which, carried on two feet as in walking, may exert a pressure on the ground of from 20 to 25 lb. per square inch. In fact many of these machines can be run safely over marsh in which a man would sink to his waist. The pressure exerted by a man when standing on the sole of one boot varies considerably, but may be taken as from 6 to 7 lb. per square inch. Some of the chain-track tractors designed for soft ground have an insistent load of less than 4 lb. per square inch, while in others, specially designed for swampy country and fitted with tracks of abnormal width, this figure falls as low as 1.8 lb. per square inch. The difference between wheel loading and chain-track loading of soft ground is shown in Fig. 37 (page 23).

*See Historical Note, page 67.

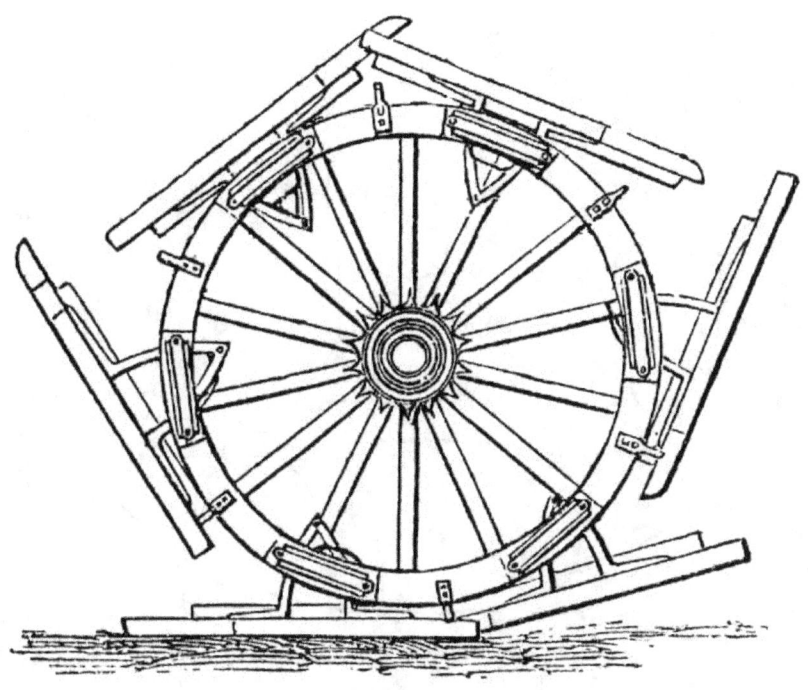

FIG. 38 – *Boydell Girdle.**

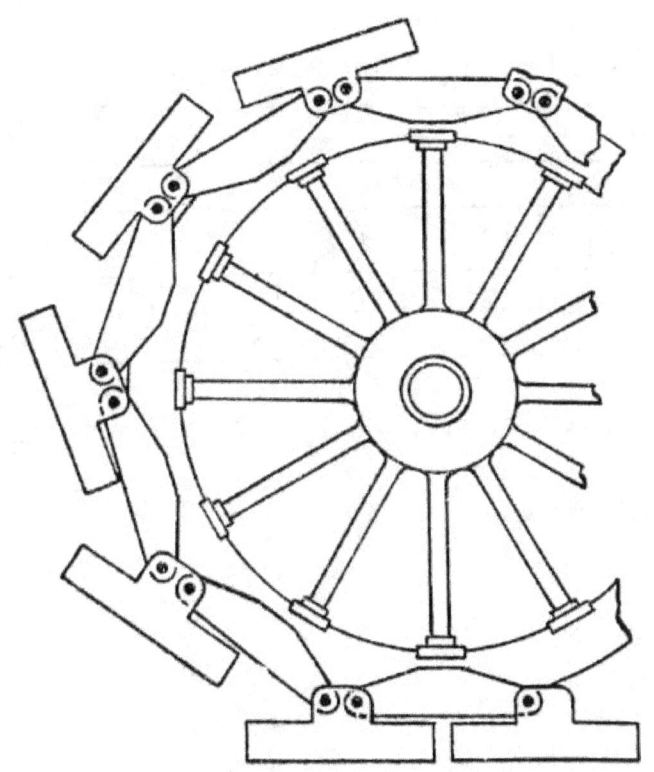

FIG. 39. - *Heavy Transport Girdle.* 1914. 6 ft. diameter over plates.

* From 'Steam on Common Roads,' 1860.

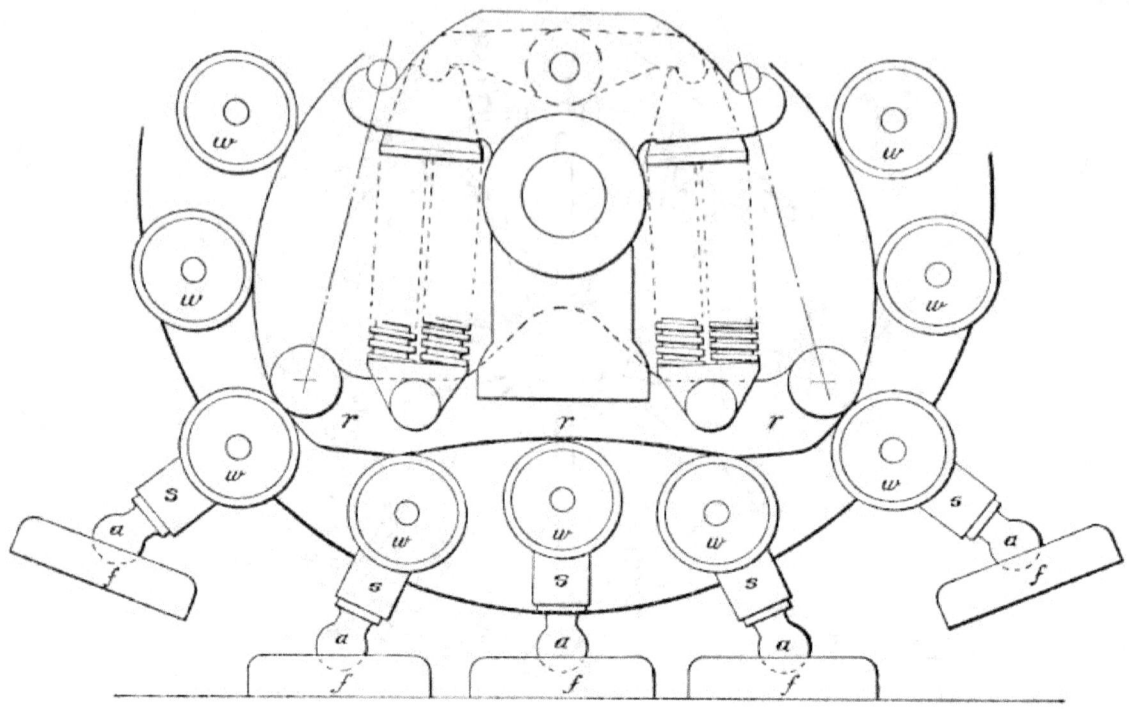

FIG. 40. - *Pedrail; Early Form* (Diplock).

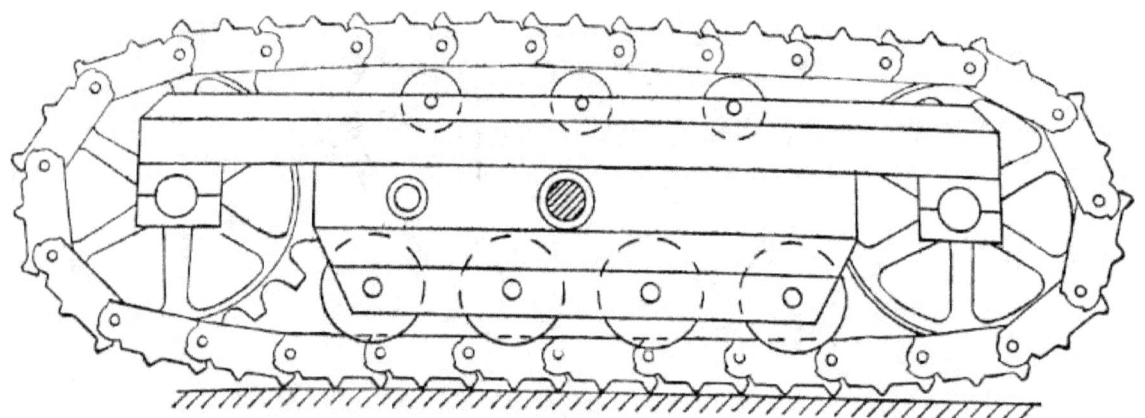

FIG. 41. - *Supporting Wheels with Aces faced Relatively to Truck-frames* (Creeping-grip).

CLASSIFICATION OF CHAIN-TRACK TRACTORS.

The various track systems may be classified as follows:-

(a) Feet or sections laid by a wheel; the girdle form, of which Boydell's tractor, Fig. 38, the ordinary girdle, Fig. 39, and the early Pedrail,* Fig. 40, are examples.
(b) Chain-track supporting wheels with axes fixed relatively to the truck-frame by which they are carried; examples of this are the Caterpillar (Holt), Fig. 47 (page 28), Creeping-Grip (Bullock), Fig. 41, Tracklayer (C. L. Best), Clayton, Strait, Burford-Cleveland, and others.
(c) Chain - track supporting through secondary chains of travelling rollers arranged as intermediate chains between the truck or sledge, and the track- or lag-chain; examples of this are the Log-Hauler (Phoenix) steam tractor for snow and ice, Fig. 42, also Fig. 43, Plate 8, the Centiped (Phoenix) petrol tractor for soft ground, Fig. 70, Plate 12, the Allis-Chalmers, Fig. 73 (page 44), and the modern form of Pedrail (Diplock) Fig. 54 (page 37).
(d) Chain-track supporting by rows of travelling balls, as in the Ball-Tread (Yuba) tractor, Fig. 46.

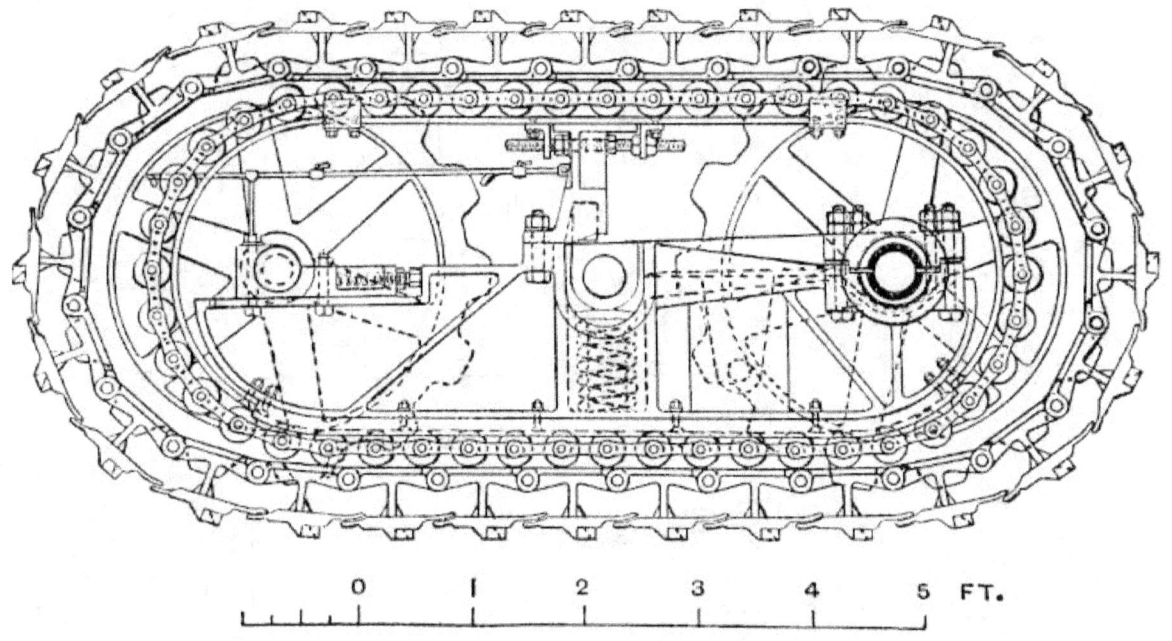

FIG. 42. - *Chain-track with Secondary Chain of Travelling Rollers* (Centiped).

Apart from the girdle, sometimes fitted to hauled vehicles, there is no modern representative of extended application of tractors of class *a* known to the Author.

The chain-track tractors that are constructed and used in the largest numbers, both as to makes and output, are those with wheels revolving on axes fixed to one or more truck-frames for each track, class *b*. It appears that chain-track tractors for the haulage of lumber were made and used commercially in the United States as early as 1904, and some of these early tractors are stated to have been still at work last year; these vehicles are fitted with the intermediate roller chains of class *c*.

Tractors in which the load is transferred from the chain to the track-frame by rows of travelling balls, class *d*, are represented at present, so far as the Author is aware, by the single example of the Ball-Tread (Yuba).

*Proceedings, I.Mech.E., 1910, pages 1561-6.

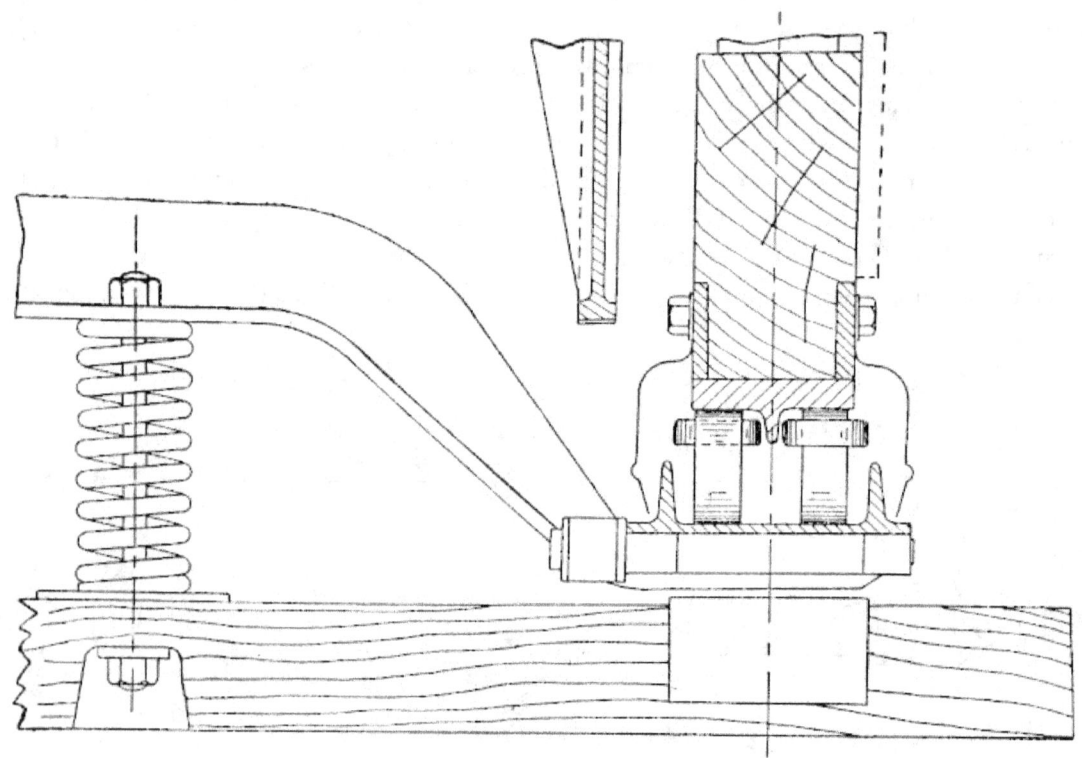

FIG. 45. - *Modern Pedrail; Foot and Rail* (Diplock).

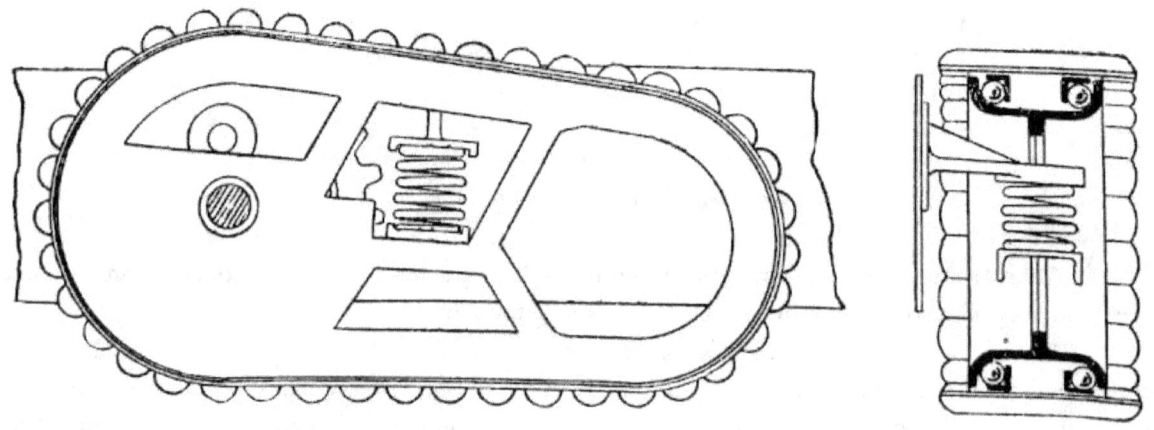

FIG. 46. - *Chain-track with, Rows of Travelling Balls* (Yuba).

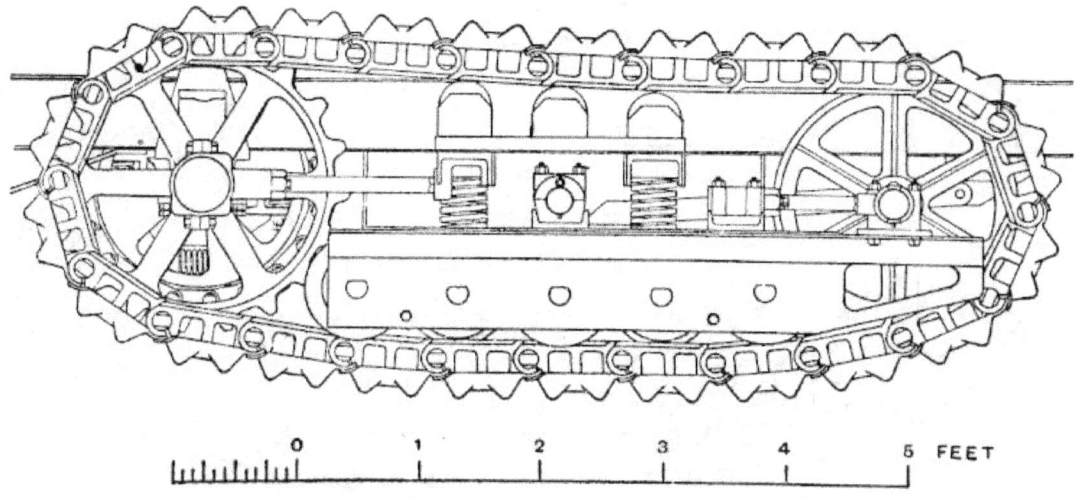

FIG. 47 - *Chain-track partly Spring-Supporting* (Caterpillar).
(For complete view, see Fig. 86, page 45.)

METHODS OF SPRINGING.

Four distinct systems of springing have been tried or are in use:-

(e) Foot springs; in this method the springs are interposed between the actual foot and the carrying portion, Fig. 45; examples are the old and new forms of Pedrail (Diplock), Fig. 40 (page 25) and Fig. 44, Plate 8.
(f) Track wholly spring supported, the axle of the track-frame being carried on springs; Centiped (Phoenix) and Allis-Chalmers, Fig. 73 (page 44).
(g) Track partly spring supported: Caterpillar (Holt), Fig. 47, and Clayton tractor (totally over effective bearing portion), Fig. 48.
(h) Track pivotally sprung; Tracklayer (Best), Fig. 49, Strait tractor (Killen-Strait), Fig. 50, and Ball-Tread (Yuba), Fig. 46 (page 27).

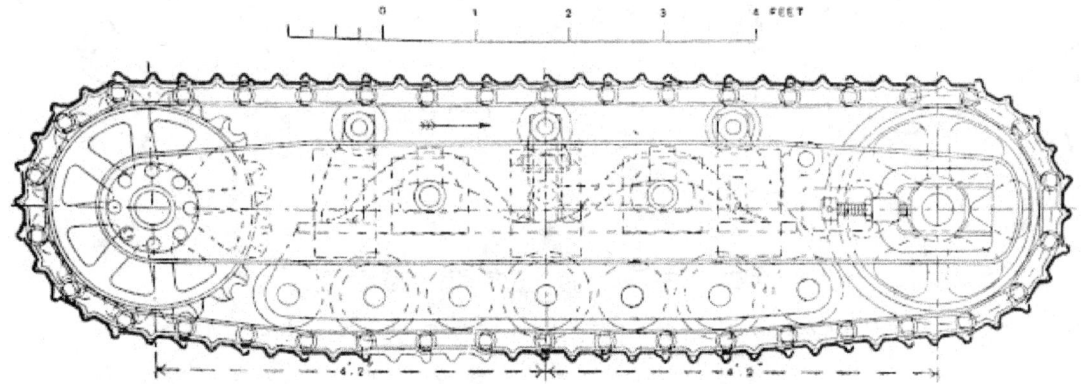

FIG. 48. - *Chain-track totally Spring-Supporting over Effective Length* (Clayton).
(For enlarged Details, see Fig. 95, page 49.)

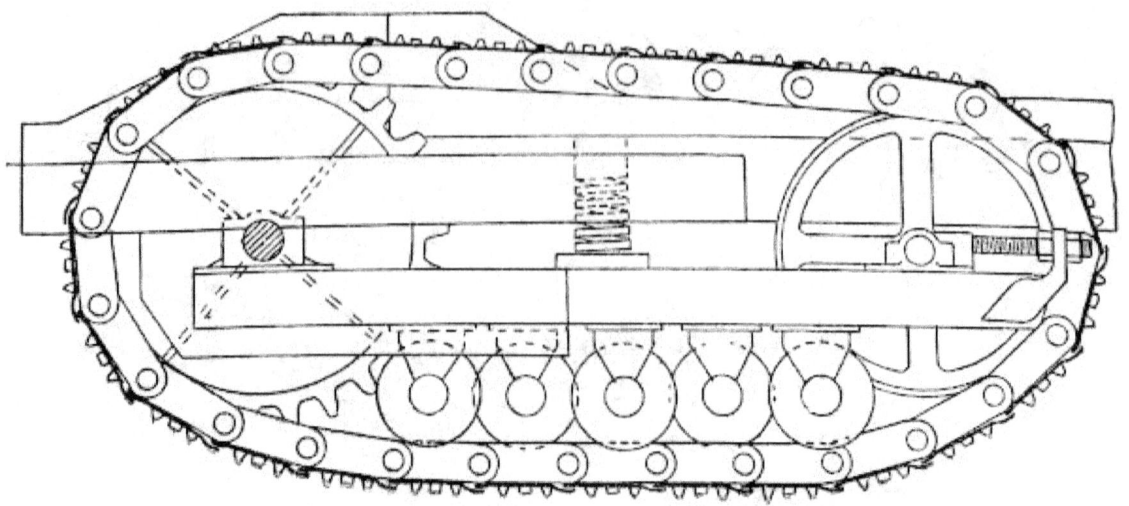

FIG. 49. - *Chain-track Frame with Pivotal-springing about Back Axle* (Tracklayer).

Owing to many constructional difficulties, direct springing of the feet, class *e*, has not yet taken a prominent place among commercial chain-track tractors. The methods of track-springing are intimately associated with those adopted for ensuring equable load distribution on the chain.

The method of supporting the vehicle on springs carried on the track-axle, class *f*, requires either a chain or a cardan drive to permit driving when the springs are unequally loaded, as when crossing ditches or running on side-lying ground.

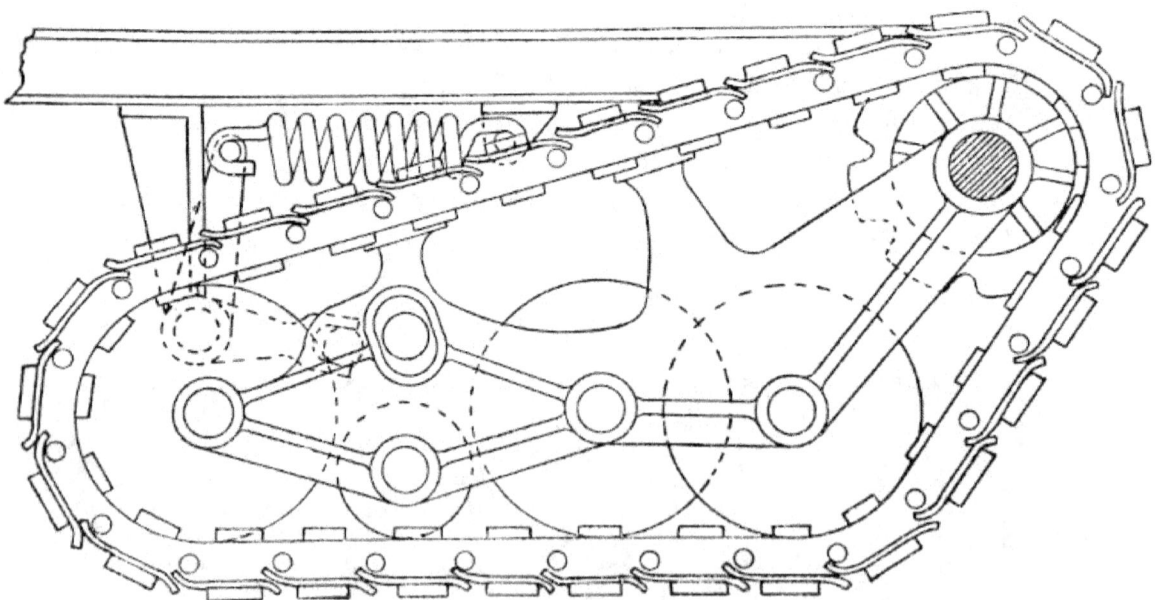

FIG. 50. - *Chain-track Frame with Pivotal springing about End Bearing* (Strait).

In some vehicles of class *g* the whole of the effective portion of the track is capable of vertical movement, as in the Clayton tractor, Fig. 48, in which a link connects the singles truck-frame to the track-frame. In others the truck is in one or more parts controlled by linkwork from the back or driving-sprocket axle, a practice followed successfully by the Caterpillar (Holt), Fig. 47.

In other tractors having chain-driven sprockets the whole of the track-frame is capable of pivoting about a central axle, an arrangement which is adopted in the Creeping-Grip tractor, Fig. 41 or the track-frame may be pivoted sprung about the axle of the driving sprocket, as in the Tracklayer (Best), Fig. 49, and the Strait tractor (Killen-Strait), Fig. 50. These two arrangements, though very different in appearance, are both equivalent to partial springing of the track, class *h*.

TRUCK-FRAMES AND TRACK-FRAME CONNEXIONS.

The various methods adopted for distributing the load of the tractor over the chain-track elements may be classified as follows:-

(j) Truck-frame and track-frame identical; this arrangement is adopted in the modern form of Pedrail, in the Strait, Fig. 50, Log-Hauler, Fig. 42, Centiped, Allis-Chalmers, Fig. 73, and Ball-Tread, Fig. 46.
(k) Truck-frame integral with track-frame; this form of construction, which very closely resembles the preceding, is adopted by the Tracklayer, Fig. 49, and Creeping-Grip, Fig. 41.
(l) Truck-frame in one piece connected by a link or guides to the track-frame; this method is used in the Clayton tractor, Fig. 48.
(m) Multiple truck-frame articulated and connected by links to the track-frame; this is adopted in the lower powered Caterpillar, Fig. 51, and enables the load on the chain-track to be distributed over irregular surfaces to which the chain-track and truck-frames can adapt themselves under the action of the springs.

When the track-frame is pivoted to the main framing, the whole length of supporting-surface of the chain-track is capable of adapting itself to long undulations or irregularities, Fig. 69, Plate 12; this arrangement is therefore very suitable for use on ordinary ground, snow, ice, sand or marsh. The articulated truck-frame, on the other hand, permits a more even distribution of load over the chain-track elements when passing over small irregularities such as stones, rocks, railway ties, rails, etc., and for this reason is better adapted for haulage from into country in which rock crops out.

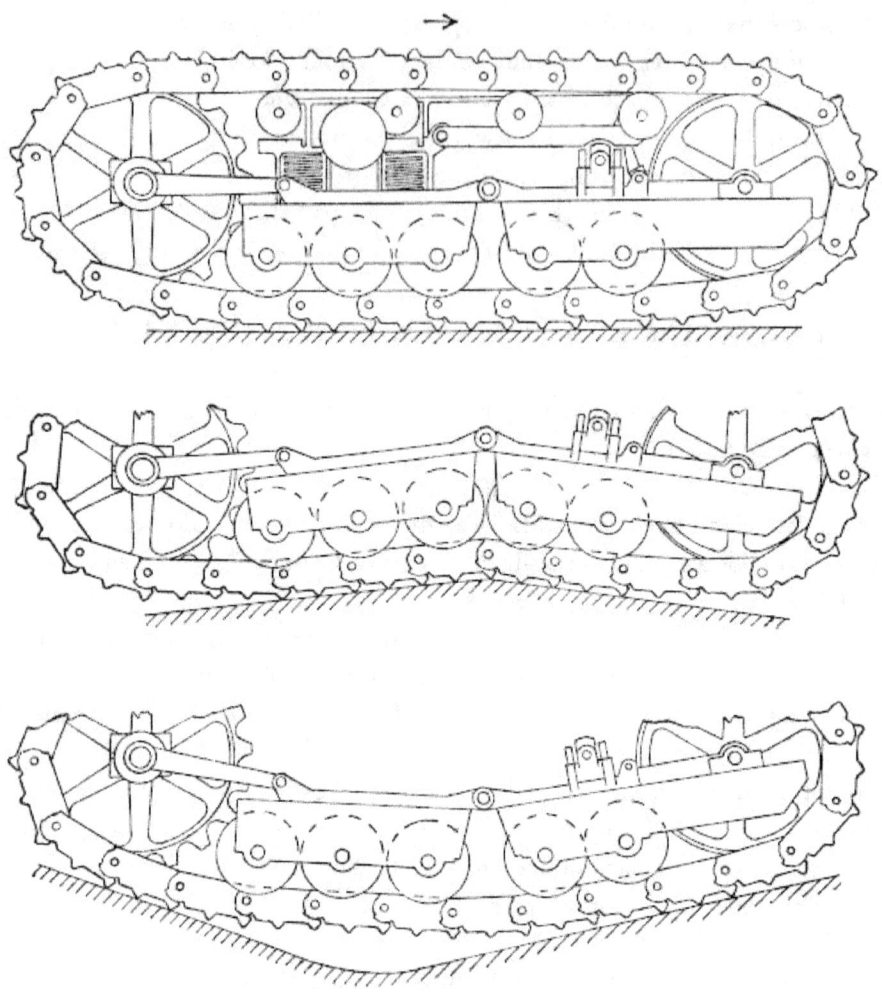

FIG. 51. - *Multiple-articulated Truck-frame* (Caterpillar).
(For complete view, see Fig. 91, page 48.)

STEERING.

The arrangements adopted for steering vary considerably. In several models, of which the Centiped and Creeping-Grip are examples, there are steering wheels with steering of the ordinary Ackermann type; in others the practice is to use a single steering wheel, as in the Caterpillar (Holt 75 h.p,), Clayton, and Ball-Tread (Yuba). The steering wheel of the Clayton is fitted with an elevating gear, so that it can be raised out of action for turning small circles. In the Strait there is a small steering chain-track; in some of the lower-powered tractors—Caterpillar (Holt 45 h.p. and 18 h.p.), Baby Creeping-Grip (16 h.p.) and F. C. Austin (35 h.p. and 15 h.p.)—the tractor is carried entirely on the two tracks, and steering is effected by releasing the clutch on the inside of the curve to be taken, and driving only on the outside of the curve. In the Burford-Cleveland a brake is applied to one side and the other is driven at increased speed through the differential gear. With some tractors, the Ball-Tread (Yuba) for example, it is possible to turn the machine about its own axis by reversing the driving direction of one of the chain-tracks.

It is obvious that there are four methods available for securing steering of tractors carried on two chain-tracks only; one by the use of the clutches alone; another by supplementing the clutch action by that of a brake acting on the appropriate side; a third by the use of the brake on one side and the action of a differential gear; and the operation of an independent reversing gear to each track gives a fourth method, In the majority of the designs under consideration it is possible to turn the machine in a circle of about 15 to 20 feet radius, and in some of the small tractors the turning circle is only 6 feet radius.

In the Log-Hauler (Phoenix) steam-driven tractor for use on snow and ice runners are fitted for steering, Fig. 59, Plate 9; these can also be arranged with some of the other types described when required to work under winter conditions.

DRIVING GEAR.

The drive of some tractors is fitted with a differential gear; this is usual where there is front-wheel steering, as in the Creeping-Grip or Centiped (Phoenix). In those machines in which steering is effected by declutching one or other of the tracks it is not usual to fit a differential; where, however, the steering is effected or assisted by independent clutching or braking of the tracks, as in the Clayton tractor and in the Burford-Cleveland, a differential gear is fitted to the main drive between the tracks.

The driving gear generally comprises a main clutch and a change-gear box of somewhat similar pattern to that used on motor lorries or trucks. The number of speeds fitted varies. In one instance, the Allis-Chalmers, there are four forward speeds and a reverse; in a few tractors three forward speeds and a reverse are fitted; two forward speeds and a reverse are more usual; very rarely, as in the high-power Caterpillar (Holt, 120 h.p.), only a single forward speed and reverse is ordinarily fitted, the two-speed gear being an alternative arrangement.

CHAIN-TRACKS.

The design of the chain-track itself is one of the most important factors of the success or failure of the tractor. The chain-track is usually driven by a sprocket at the back end, and passes round an idler at the front carried on the track-frame and engaging only laterally with the chain-track by a central flange or flanges; the idler axle is usually fitted with an adjustment for taking up wear of the chain-track. It will be noted from the cross-sections shown that the width of the track is greater than the width of bearing for the rollers, and any hard object, rock or rail, encountered by the chain may therefore tend to load it unequally, and subject it to torsional stress. Furthermore, if the chain-track passes over a hard and narrow object, such as a railway rail, this will tend to produce reverse curvature in any unsupported portion of the chain. This is particularly the case where the chain has a short free length between the leading idler-wheel and the first pair of carrying rollers. Realization and over-estimation of these difficulties were probably responsible for the early designs of chains made in this country.* The links were so arranged that the connecting-pins were raised several inches from the ground, and the blocks, of which the chain-track was composed, were so made as to butt against each other when the chain-track was in the position of minimum

In this way the bearing portion of the chain-track curvature. In this way the bearing portion of the chain-track formed an inverted arch, which was supported at portions of its length by the rollers, and depended on contact between the links for carrying the load on intermediate portions of the chain, as shown in Fig. 52. In order to reduce the stresses still further, and to relieve the chain-pins of the heavy load to which they would otherwise be subjected, interlocking extensions above the pins were formed on the chain links. Unfortunately this method of carrying the load resulted in the introduction of new and greater difficulties, generally known as nut-cracker action. If the track worked only on sand or mud, or even on pure clay, there would be no difficulty; but where there is a combination of stones and clay or when, in crossing gravel or roads, any hard object can be picked up by the track and get between the bearing faces of the links, the stresses so caused become excessive and the track may even become jammed. This result can be caused also by pieces of fencing wire or scrap iron such as may occasionally be picked up on a farm. In the forms of chain-track now commonly used the pin centres are brought as close to the ground level as possible, and the chain-plates are so formed as to shield the opening between the track-plates over the working range, as will be seen from Figs. 89 and 95 (pages 47 and 49).

The second source of trouble to which chain-tracks are subject is that of stretching due to wear. This lengthens the pitch of the chain and necessitates fitting an adjusting gear to take up the slack; it also tends to cause riding on the driving sprocket. As it is usual to pitch the chain-pins so widely that they will engage with alternate teeth of the driving wheel, when cut as an ordinary sprocket, this wheel should have an odd number of teeth in order to distribute the wear uniformly over the whole of the teeth.

The chain-track links have three functions to perform:-

(1) Carrying the load.
(2) Affording the necessary resistance for traction by engaging with the ground surface, and
(3) Providing the gearing pin, or its equivalent in tooth form, to engage with the driving sprocket.

* See Historical Note, page 67.

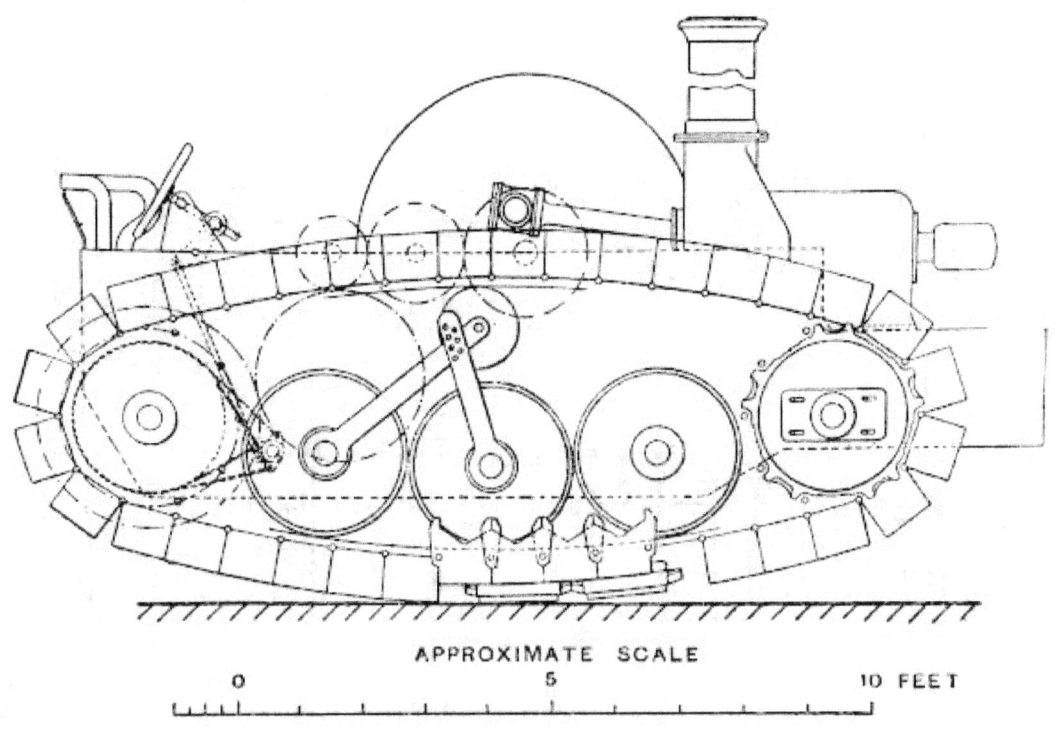

FIG. 52. - *Roberts Tractor; Side View.*
Illustrating Nut-cracker Action. 1904.

The first condition involves a machined surface, preferably hardened, for the carrying rollers to bear on. The second condition necessitates an irregular outline of the lower surface of the link, so that it can gear with the ground surface as a large pinion would gear into a rack; hence the form of shoe must approximate to a ribbed section, and is usually of pressed steel. The third condition requires that the pin-joints or gear faces shall be machined accurately enough to work smoothly with the driving sprocket without risk of riding on its teeth. These three conditions, coupled with the proximity of the connecting pin to the ground, have limited the best designs to built-up forms of construction in which the actual link is a machined steel-casting riveted to a pressed steel sole-plate.

The conditions of running require that this plate shall be of such form as to clear the end of the next plate throughout the range of inclination to which the links are subjected, and that over the same range there shall be no opening between the plates, in which nut-cracker action can take place (compare Figs. 48 and 52). The troubles due to stretching can be forestalled, as was shown by Hans Renold about twenty-five years ago, by making the pitch of the driving-wheel large in the first instance, so that the chain-track is driven, at the beginning of its life, from a tooth nearly opposite that which is entering into engagement with it, Fig. 53.

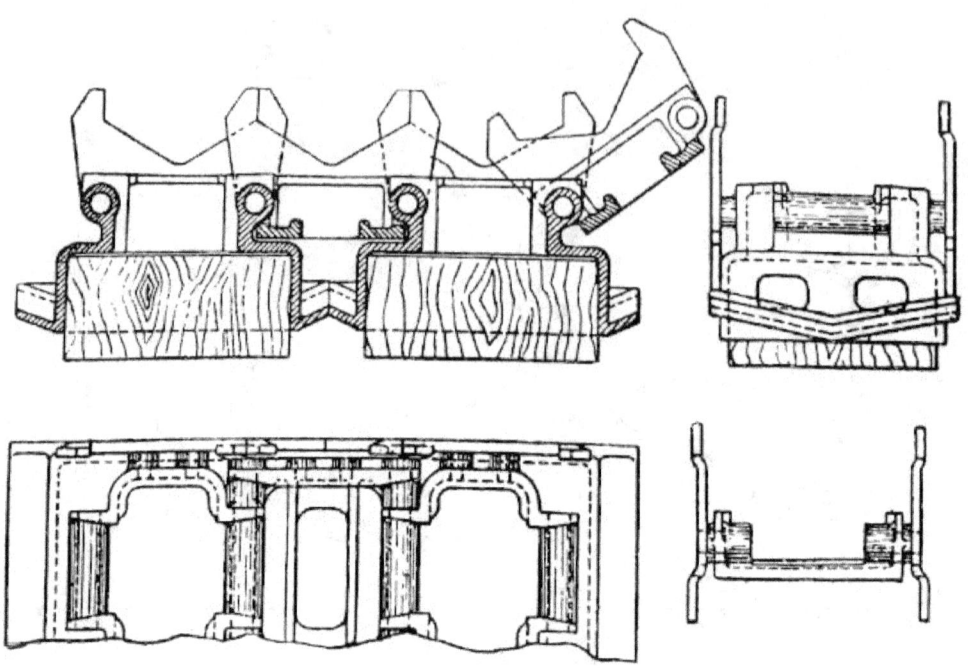

Detail of Chain-track; Section.

New – Chain pitch less than wheel pitch.   **Worn –** Chain pitch equal to wheel pitch.

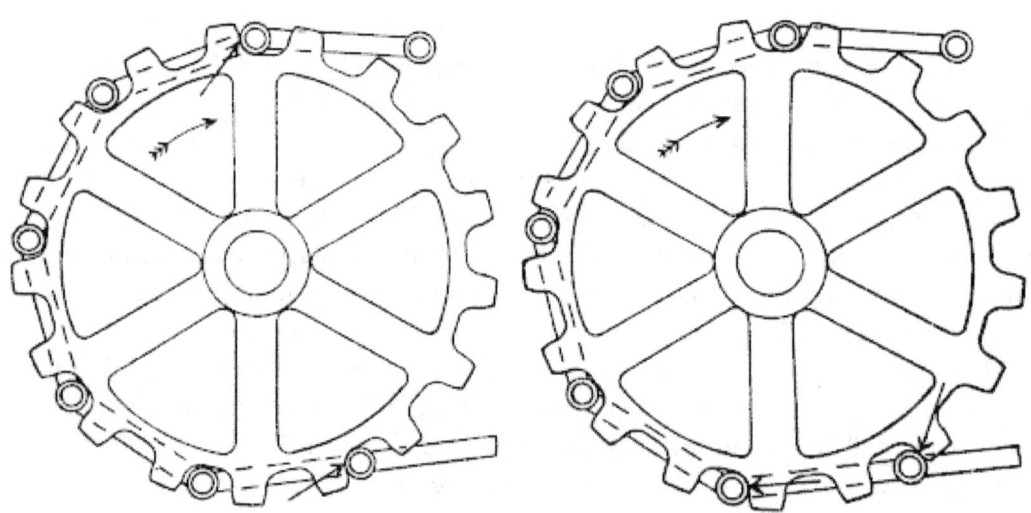

FIG. 53. - *Sprocket Design to Allow for Wear of Chain* (Hans Renold).

By these means a fairly long life can be assured to the chain, provided it is not overloaded, and that efficient lubrication is maintained. The usual method is to supply a large amount of low-grade oil to the inside of the chain; an oil drip so fitted to the side chains of certain early pattern motor-cars (1900 - 1902), which ran under almost identical conditions of mud and dust, has proved almost as efficient in preserving the chains as frequent removal, washing in paraffin, drying, and soaking in a bath of warm tallow and plumbago. Cheap oil used lavishly appears, in fact, to keep the tracks from cutting or seizing even under very bad conditions, In the Tracklayer (Best) a different method of providing against wear is adopted, the pin being made of rocker form, Fig. 55, Plate 8.

Another matter that affects the design of the whole chain-track is that of the permissible sag. The chain is supported laterally by the sprocket teeth at the back end and by the idler-rim at the front end; over the intermediate parts it is only supported by the flanges of the supporting rollers. As the flanges of these rollers are limited for size by the closeness of the bearing surface to the ground, it will be seen that the permissible sag which may occur when crossing holes in rough ground is not very great. If supported at the ends the suspended portion of chain hanging freely will assume a curve which approximates to a catenary. For such a crude approximation as is afforded by the few suspended links of the chain-track, the curve passing through the centres of the pins is very nearly a parabola; if the curvature is small, this in turn approaches a circular arc. The worst condition for the chain occurs when there is a combination of steering and backing on rough ground; this may at the same time cause slackness in the carrying part of the chain combined with sideway thrust due to steering, and, in some of the forms of construction described, the roller flanges may be caused to ride on the chain surfaces.

One method of dealing with the difficulty is that of giving a large camber to the supporting portion of the track. In a vehicle styled the 'Centipede' built by Wm. Foster and Co., Ltd.,* the track was given a camber about equivalent to that of a 16 feet wheel. In the majority of the tractors described the curvature of the supporting portion of the chain-track is very small, corresponding to that of wheels from 90 feet to 160 feet in diameter, and in some cases even it has zero value, the part of the chain-track on which the wheels run being laid flat. This necessitates care in keeping the chain-track constantly adjusted for wear. It will be noted that this difficulty is not so serious in the chain-tracks fitted with an intermediate chain, as both the chains can sag together, and it is possible to provide deep retaining faces to keep the intermediate roller-chains in place. The lateral pressure on the chain-track when the tractor is travelling on side-lying ground, or crossing ditches obliquely, is another factor that must be taken into account in designing the chain-track as well as the truck-frame and its connexions. Under most conditions the sideway thrust is taken partly on the flanges of the truck wheels and partly on the flanges of the driving-sprocket and idler. An arrangement which appears to provide specially for this condition is that of the Ball-Tread (Yuba), Fig. 46 (page 27). In this the whole of the lateral as well as the vertical pressure is transferred from the chain-track to the track-frame through the balls. Lateral load merely affects the angle of incidence of pressure on the ball without increase of friction. Diplock, in his recent developments of the Pedrail, has obtained the same result in a different way, by using a compound chain of rollers arranged alternately with axes horizontal and vertical, the track-frame and bearing surfaces of the chain-track- being made with appropriate faces for receiving the vertical and lateral loads respectively, Fig. 54.

* The Implement and Machinery Review, 1st November, 1914, page 888. The text describes the vehicle as being "carried on twenty turned steel wheels 16 inches in diameter," but the illustration shows only four axles for carrying wheels, corresponding to sixteen wheels. Moreover the power is given as 60 b.h.p. at 1,000 revolutions per minute, the low speed as 2 miles per hour, and the drawbar pull only 4,000 lb. which correspond to an overall mechanical efficiency of about 35.5 per cent.

The radius of curvature of the track of the experimental tractors made in England about 1904 was also about 16 feet).

TRACK SHOES.

For travelling on rough country and under general agricultural conditions the pressed steel shoe is found satisfactory, Fig. 56, Plate 8. In this case the supporting surface for the rollers consists of a steel casting, or case-hardened drop-forging, bolted to the shoe; in other cases the track shoe and supporting surface are formed in one as a single steel-casting, Fig. 57, Plate 8.

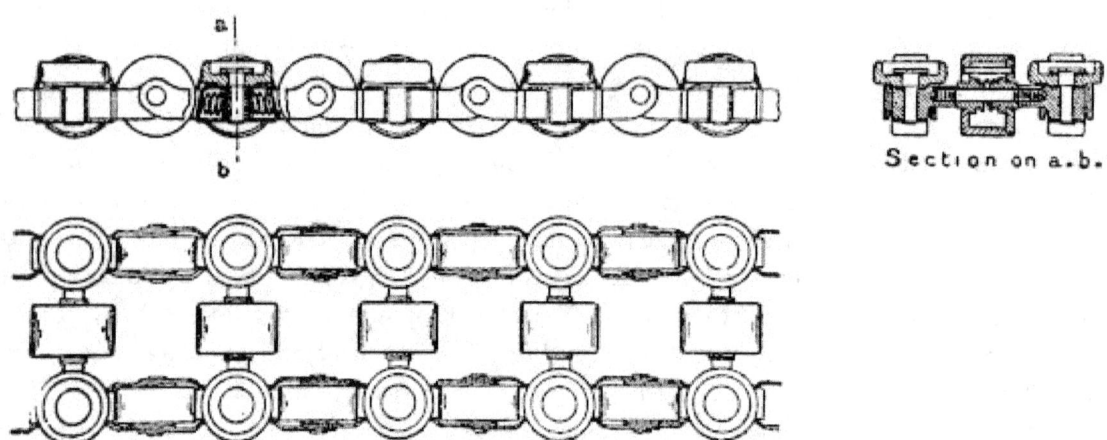

FIG. 54. - *Roller-chain for Pedrail* (Diplock).

Where the ground surface presents small resistance to shear, or actually lubricates the shoes, as is the case with clay and chalk and some grass surfaces, it is sometimes found necessary to attach grousers or spuds of angle-bar to the shoes by means of bolts. When the tractor is required to do hauling on ordinary roads it may be necessary to attach thin sleepers of hard wood to the shoes in order to provide a flat bearing and prevent damage to the road surface. Those chain-tracks which present fairly flat surfaces to the road and excessively cambered - the lower-powered Caterpillars, Clayton, and Burford-Cleveland for example - can be run over inferior roads with even less risk of damage than would occur with the ordinary driving slats of the common traction engine.

ENGINES.

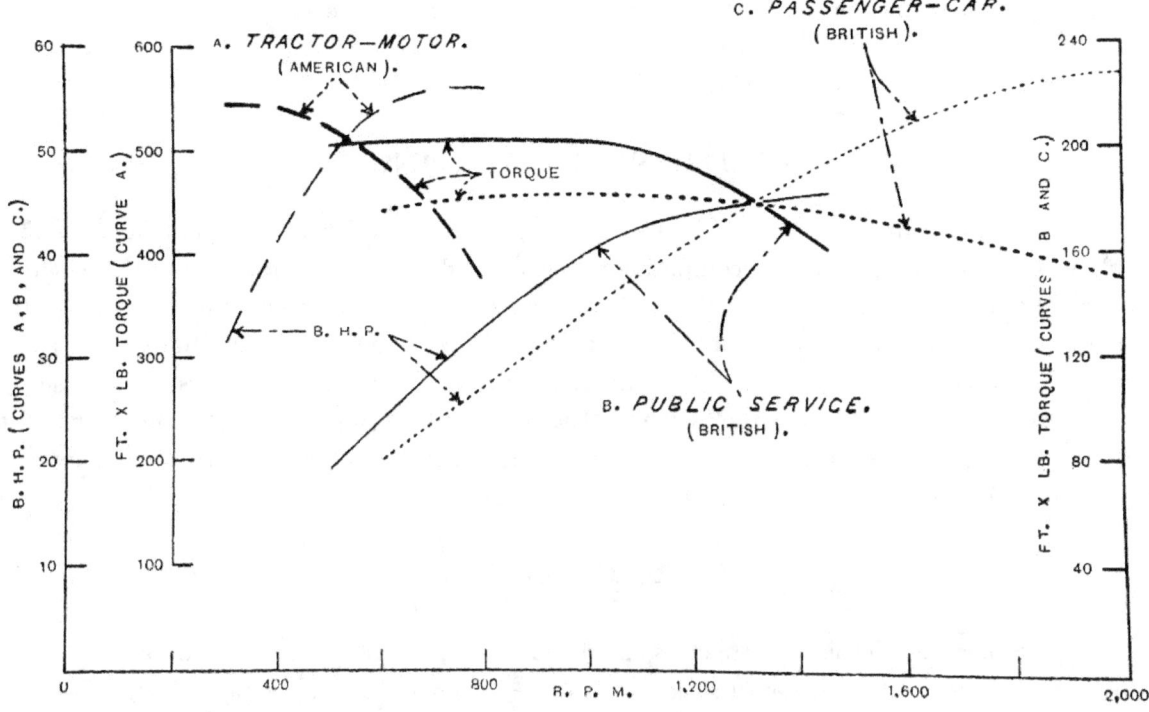

FIG. 58. - *Torque and Horse-Power Curves for Motors.*

The internal-combustion engines fitted to chain-track tractors are generally of a type resembling that used for heavy lorry or truck work. There is no reason for great economy in weight, and indeed the weight of the vehicle must not be too small for the tractive effort; consequently cast-iron is preferred to aluminium for the engine and also for the gear-box.

A governor is usually fitted to prevent racing the engine. That care which is taken in designing the passenger-car engine to obtain a flat torque curve, and which is accompanied by large diameters of valves, induction and exhaust passages, is not called for here. The torque curve consequently falls rapidly at a comparatively low speed. This, as is generally recognized by the makers of engines for commercial vehicles, is not without advantage, for when the resistance to traction becomes unduly heavy, owing to gradient or softness of the ground, and the speed of the engine is consequently reduced, the torque increases and the driver has a better chance of extricating the vehicle from its difficulty without changing gear than is the case with constant torque, A typical torque curve of one of these engines is given in Fig. 58, The tractor engine runs at a low piston speed, 600 to 900 feet per minute ; that is, at barely one-half the speed usual for the passenger-car engine, On the other hand, it is required to work at a greater percentage of its full power than is the case with the passenger-car engine, and for considerations of fuel economy it should, in general, be designed to give its best efficiency, measured in b.h.p. per gallon per hour, at from 50 to 70 per cent, of its full rated capacity. In many instances the engines are designed to start on petrol (gasolene) and to run normally on paraffin, or, in America, on 'distillate'; * this also implies a larger engine for the power than is the case where petrol or benzole is the usual fuel.

Lubrication of the engine is usually ensured by the splash system, but in some instances a combined splash and forced lubrication system is fitted. Separate oil reservoirs are of course required for the engine-lubricating oil and for the crude black oil used for the track-chains.

* Distillate' is a low-grade spirit about 0.770 to 0.780 density.

IGNITION AND STARTING.

The usual form of ignition is the high-tension magneto. Self-starters are not fitted, as they would involve too much additional complication. Owing to the large size of some of these engines, and the difficulty of starting by swinging the starting crank-handle, a trip-gear is arranged on the magneto which enables a strong sparking current to be obtained for a small movement of the cranking handle. This arrangement was used by the Author in 1895 on gas-traction tramcars, the engines of which were started by pulling over the fly-wheel. The combined use of petrol for starting and paraffin for running complicates the carburettor and piping arrangements slightly, and involves fitting an extra tank, but the advantage of ease in starting is well worth this.

RADIATORS.

The radiators of chain-track tractors have been designed to meet the exigencies of the working conditions. The cooling surface must be large to enable the engine to run continuously under heavy load in the heat of summer; and a fan, if fitted, must be easy of access and simple in its driving gear. The water-supply carried must be sufficiently large to last for a long journey of perhaps a whole day away from the base station; for this reason a large tank for water, or equivalent radiator capacity, is provided in most models. Finally, it is necessary that the whole water system should be capable of being drained at its lowest points, to avoid risk of damage by frost when the machine is laid up in winter.

As the machines have to work in places distant from large manufacturing centres, it is moreover advisable that the design of radiator should admit of its easy removal from the vehicle, of ready access to its water joints, and of the possibility of executing repairs without special tools and appliances. The conditions of the problem are similar to those of the motor omnibus; more severe regarding the total cooling to be effected, and less severe in weight restrictions. The type generally adopted is the vertical gilled-tube with top and bottom reservoirs; in some of the lower-power tractors these reservoirs are made of sufficient size to avoid fitting a separate water tank. The water circulation is usually forced by means of a centrifugal pump or by a gear pump, the former type being the more common.

SPEED OF TRACTORS.

At normal revolutions per minute of the engine the lowest or first speed of the majority of chain-track tractors varies from 1.25 to 2.25 miles per hour; on the second speed from 2 to 3.5 miles per hour, and on the third or top speed? where such is fitted, from 3 to 5.5 miles per hour,.

The changing of gear is usually effected by sliding gears or dog-clutches in a gear-box of motor-lorry type, the main clutch being disengaged for gear-changing or reversing. Cut steel gears, case-hardened, are usual for spurs and bevels; plain bearings are commonly fitted in the gear-boxes, which are of heavy design and usually of cast-iron; ball-bearings are only used in a few instances. In the Caterpillar and Creeping-Grip tractors transmission to the back axles is by chain-drive; in the Tracklayer and Ball-Tread it is by spur-pinions gearing into an internal gear-wheel formed in one with the driving sprocket.

It is curious that in these heavy slow-moving vehicles the epicyclic form of gear-change, usual on American passenger cars, should have found no practical application. This is the more remarkable as the difficulty of gear-changing with sliding gears or dog-clutches is greater with governed engines.

DRAWBAR PULL.

Chain-track tractor makers are nearly unanimous in their estimates of the loss of power that takes place between the engine and the ground. The power delivered at the track is estimated at from 70 per cent in the large 100 h.p. tractors down to 55 per cent in the small tractors of 30 h.p. and under. The force required to haul the track itself apparently varies from 15 per cent of the weight carried in the fixed roller type with plain bearings, to 2 per cent of the weight carried in the travelling roller type of track, or its equivalent.

The difference between the two constructions in loss of tractive effort may be attributed in part to the fact that it is easier to ensure efficient lubrication in a roller system that is loaded intermittently, than in one which is subjected to continuous load in one direction.

The comparatively heavy resistance to traction in those chain-track tractors in which the load is carried through rollers running on fixed axles is reduced in some cases by fitting roller bearings, of the Hyatt pattern for instance, on the roller-wheel shafts. These do not, however, reduce the lateral friction on the flanges and bosses of these wheels, which depends on the supply of grease that can be forced in through the pins by lubricators of the Stauffer type. Roller bearings so fitted will, of course, reduce the starting effort; nevertheless, it appears to the Author that, of the several sources of friction contributing to losses in the chain-tracks, the lateral friction of parts of the system of carrying rollers, where the rollers revolve on fixed axes, probably accounts for more than half the total loss of tractive effort.

Experience, gained in comparative tests of the fixed-axles and roller-chain systems made over a long period under the same working conditions, will be necessary to determine the cost of upkeep of each; the same tests will also show how far the extra complication of the intermediate travelling chain of rollers is compensated by reduction in fuel consumption.

CLIMBING POWER.

The drawbar pull varies in different types of vehicle from 30 per cent to over 70 per cent of the weight of the vehicle; a heavy tractor weighing 28,000 lb. can exert a drawbar pull of 12,000 Ibo; a small tractor of 30 h.p. can exert a pull of 3,500 lb. The steepness of gradient that these vehicles can climb is consequently very great.

On railways gradients are measured as the tangent of the angle of inclination, the rise being referred to the plan of the track. In the case of hill-climbing tests of automobiles the maximum gradient climbed rarely exceeded 1 in 3; for this value of the tangent, the angle with the horizontal is 18° 26', of which the sine is 0.3162; that is to say, the actual length of the road, in elevation, is about 5 per cent more than that given in plan. As some chain-track tractors are capable of climbing a gradient as steep as 45°, it is necessary to refer climbing and haulage problems to the actual road length, which for an inclination of 45° is over 40 per cent greater than the length in plan. Thus a vehicle capable of exerting a draw-bar pull equal to 50 per cent of its weight can climb a slope of 30° provided that the ground presents sufficient adhesion; and a vehicle capable of exerting a drawbar pull equal to 71 per cent of its weight can climb a slope of 45° provided that a sufficient bite can be obtained by the tracks on the surface.

The tractive effort for ordinary railway practice speeds not exceeding 5 miles per hour, and therefore free from questions of air resistance, varies between 9 lb. per ton (0.4 per cent) for the train hauled, and 12.5 lb. per ton (0.55 per cent) for the train including engine and tender; on tramway rails the tractive effort varies from 30 lb. per ton (1.34 per cent) wet, to 56 lb. per ton (2.5 per cent) dry; or ordinary roads the effort varies from 70 lb. per ton (3.1 per cent) on macadam to 120 lb. per ton (5.4 per cent) on good gravel roads, and it is much more on bad gravel roads and across fields. The inclination of the ground on which chain-track tractors will roll back varies from 1 in 7 for the fixed-axle plain-bearing roller type, to 1 in 50 for the independent roller-chain or its equivalent.

In comparing chain-track tractors with other vehicles for resistance to traction it must, however, be remembered that the resistance with fixed axle-wheels, which may amount to 14 per cent for the tractor itself with its gear, or to 9 per cent for a trailing tractor, is not greatly increased by conditions under which a wheeled vehicle sinks so deeply that the tractive effort required to haul it may amount to 40 per cent of its weight. Conditions produced by swamp or sand can be such that no horse or wheeled tractor could haul the load at all, yet they will present no difficulty to the chain-track tractor.

DRAWBAR CONNEXION.

In vehicles exerting so heavy a drawbar pull as is possible with the chain-track tractors, it is of great importance that the drawbar connexion should be placed as low as practicable without unduly reducing the ground clearance. The couple formed by the drawbar pull at the height of the drawbar, and the ground resistance, acts on the frame of the tractor in the same direction as the couple due to the tractive effort applied at the sprocket teeth; both tend to cause the front of the tractor to rise, and, as will be seen from Table 2 (pages 78 to 83), the portion of the load carried on the front wheel is usually small. For this reason it is advisable in designing vehicles of short base and heavy drawbar pull that the drawbar shall be placed low. In vehicles intended for soft and marshy ground it is desirable to use wider treads rather than to attempt to obtain very great ground clearance for the drawbar. For this reason many makes are fitted with alternative widths of chain-track according to the land on which they are to work.

LEADING FEATURES OF TRACTORS.

The horse-power, engine dimensions and speed, tractor speeds forward and reverse, track width, length of the track in contact with the ground, weight on track, weight on steering wheels (if any), load per square inch of track, dimensions of steering wheels, radius of turning circle, pitch of chain, capacity of petrol, water and oil tanks, drawbar pull, and ratio of drawbar pull to weight of tractor are given in Table 2 (pages 78 to 83).

Some further particulars of the characteristic features of the several types may now be given.

THE LOG-HAULER, PHOENIX.

The Log-Hauler (Phoenix) was first used in 1904, and engines of this class were supplied to Montana in 1906; many are in use in Wisconsin. One of these vehicles is shown in Fig. 59, Plate 9. In early logging a team of four bullocks hauled a sleigh carrying about 1,500 feet of timber, or a weight of about 8 tons. The present day sleigh is from 7 to 8 feet wide from centre to centre of runners, with cross bunks, 12 to 16 feet apart, on which loads are built up, Fig. 62, Plate 10; the load for an individual sleigh amounts to from 5,000 to 7,000 feet, and the train consists of from seven to fifteen sleighs, or a weight of 200 to 420 tons, Fig. 64, Plate 10.

The Log-Hauler has a locomotive-type multitubular boiler with barrel 15 feet long, 36 inches diameter, with 1.75-inch tubes, the working pressure being 200 lb. per square inch. Coal or wood fuel is used. The engines are vertical with two double-acting cylinders, 6.25-inch bore, 8-inch stroke, with link-motion reversing gear on each side, Fig. 60, Plate 9. Each drives through spur-gearing into a lay-shaft which itself drives at the back through a bevel-pinion into a bevel-wheel fixed to a spur wheel which is free to revolve on the back axle, This spur wheel gears through an intermediate spur wheel into the rear sprocket, Fig. 60, Plate 9; the front sprocket is an idler. The back sprocket gears into the outer or lag-chain of shoes, which is the track-chain; the centres of the pins are about 2 inches from the face of the shoe, which is fitted with a Λ projection to grip the frozen surface on which the Log-Hauler travels. There is no special provision in this case for guarding the joints of the chain-track. A screw adjustment is provided for taking up stretch of the lag-chain by acting on the front sprocket bearings. There are two intermediate roller chains with case-hardened rollers 3 inches diameter and about 2 inches wide for each track; these roller chains run on tool-steel guides secured to the underside of the truck frame, and are guided at the sides where they pass over the carrying segments. Lateral pressure is taken on the sprockets. The speed of running is from 4 to 5 miles per hour. The load hauled depends not only on the grading of the road but on how well it has been iced. The crew consists of the engineer (driver), a fireman, and the pilot; the latter sits in front of the smoke-box, so that he can see both the road and the runners, Fig. 63, Plate 10. The steering column is inclined backwards, and drives through a worm into a cross shaft which is connected by chains to the ends of the cross beam of the runners. The coal consumption is stated to be from 56 to 84 lb. per mile. The engines require special housing arrangements, similar to those provided for locomotives in Canada or Russia, as they have to work in country where the night temperature may fall as low as -50° F. The tractors have to haul their own water supply, and may also be required to plough the track clear of snow. Provision is made for attaching a plough clear of the runners.

The illustrations, Figs. 62 - 65, Plates 10 and 11, show the methods of building up the loads and trains of sledges both for log and sawn timber; they also show the small ground clearance allowed in this particular type of tractor, and the curve that can be taken by the train.

The Centiped, Fig. 66, Plate 11, resembles the Log-Hauler in the pivotal arrangement of the track frames, and in the construction of the lag-chains and roller-chains. Both axles are carried on springs. The steering is of the Ackermann type with the front axle pivoted about a horizontal axis, so that the vehicle has virtually a 3-point suspension, and either front wheel can be taken over an obstacle 10 inches high. This is a very necessary feature in tractors fitted with two front wheels, as they are frequently required to cross ditches obliquely, Fig. 68, Plate 12, or banks of railway tracks, Fig. 69, Plate 12, Examples of the Centiped at work are given in Figs. 71 and 72, Plate 13. Power is transmitted through a three-speed and reverse gear-box to the differential shaft, and thence to the rear sprockets in a manner similar to that adopted in the Log-Hauler.

The track adjustments, Fig. 70, Plate 12, are of the same type as on the Log-Hauler, but the construction of the chain-track is different. The centres of the pins are arranged unusually far from the face of the shoes, this dimension being about 5 inches; but to avoid nut-cracker action each shoe overlaps the one in front by about 2.5 inches on the flat and 1 inch at maximum curvature. Each shoe carries a strake about 2 inches wide by 1 inch thick. The small amount of sinking of the track on soft sand, as compared with that of the wheels, is shown in Fig. 67, Plate 11.

THE ALLIS-CHALMERS TRACTOR-TRUCK.

This vehicle, Fig. 73, closely resembles the Centiped in general arrangement of gears as well as in the design of the chain-tracks and of the roller chains. It is fitted with an engine of higher speed, of 68 b.h.p., and is peculiar in having four forward speeds, the third speed on direct drive giving about 6.2 miles per hour, and the indirect-driven fourth as high as 7.5 miles per hour. The springing of the front axle is effected by two heavy coil springs; the back axle is carried on two side springs of the semi-elliptic sliding type, each 54 inches long by 3 inches wide, and supplemented by a cross-spring, 30 inches long by 3 inches wide. The front-wheels run on roller bearings. A special feature of the design is the large road clearance permitting the vehicle to be used over very rough ground or in deep snow. The Allis Chalmers Tractor weighs about half a ton more than the Centiped, but has approximately the same insistent weight.

FIG. 73. - *Tractor-truck* (Allis-Chalmers).

THE CATERPILLAR, HOLT.

The load is carried in the Caterpillar (Holt) from the central portion of the chain on five pairs of chilled cast-iron flanged wheels. The wheels run on heat-treated steel axles fitted with Hyatt roller bearings. The pairs of wheel flanges are alternately inside and outside the bearing surfaces of the track-chain. The track-links are either of annealed steel castings, as in the 120 h.p. model, Fig. 57, Plate 8; or are built up of steel drop-forgings machined on the upper face, Fig. 56, Plate 8, and secured to pressed steel sole plates ; the shoes have each two corrugations about 1.5 inch wide for gripping the ground; the pins are of case-hardened steel. The bearing faces are 2.5 inches wide in the 75 h.p. and 2.25 inches in the 45 h.p. model. As there are two lines of rail surfaces on each chain, the total effective rail-width is 10 inches for the 75 h.p. and 9 inches for the 45 h.p. models respectively. The truck in both the 120 h.p. model, Fig. 74, Plate 13, and in the 75 h.p. model, Fig. 75, Plate 14, and Figs. 86—88, is in a single piece, and is connected to the axles by two links, of which the front one is fitted with a screw adjustment but is allowed freedom for vertical movement relatively to the track-frame; the load is carried by four double-coil helical springs on each side. The unsupported portion of the chain-track between the sprocket or idler and the adjacent wheel at each end is about 1½ links in length. The steering is effected by a single front wheel controlled by a non-reversible worm and wheel gear.

Power is transmitted from the engine through a multiple-disk clutch with five plates, bronze to cast iron, to simple reversing gear-box with two speeds and reverse; the drive-shaft is spring-driven to reduce shock in taking up the load; the track sprockets are driven by chains; there is no differential gear, but friction clutches are fitted on both sides so that the drive may be on either side to facilitate turning. Examples of Caterpillars at work are given in Figs. 81-85, Plates 16 and 17.

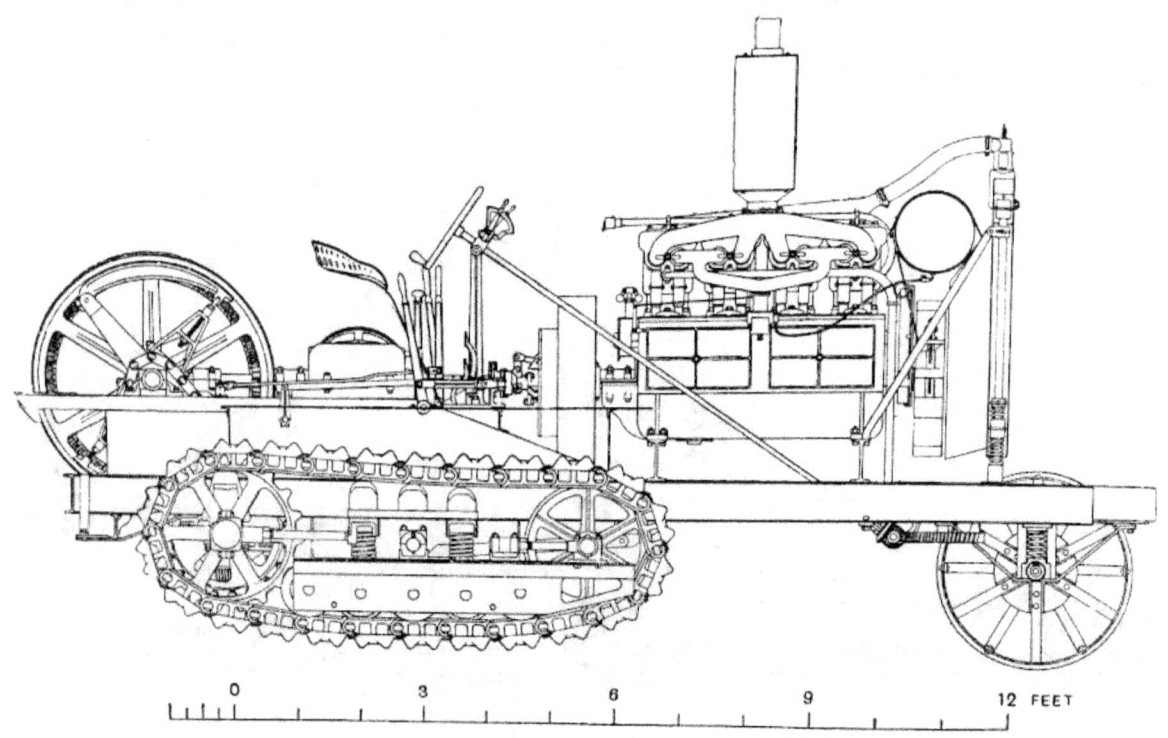

FIG. 86. - *Caterpillar; 75 h.p., Side Elevation.*

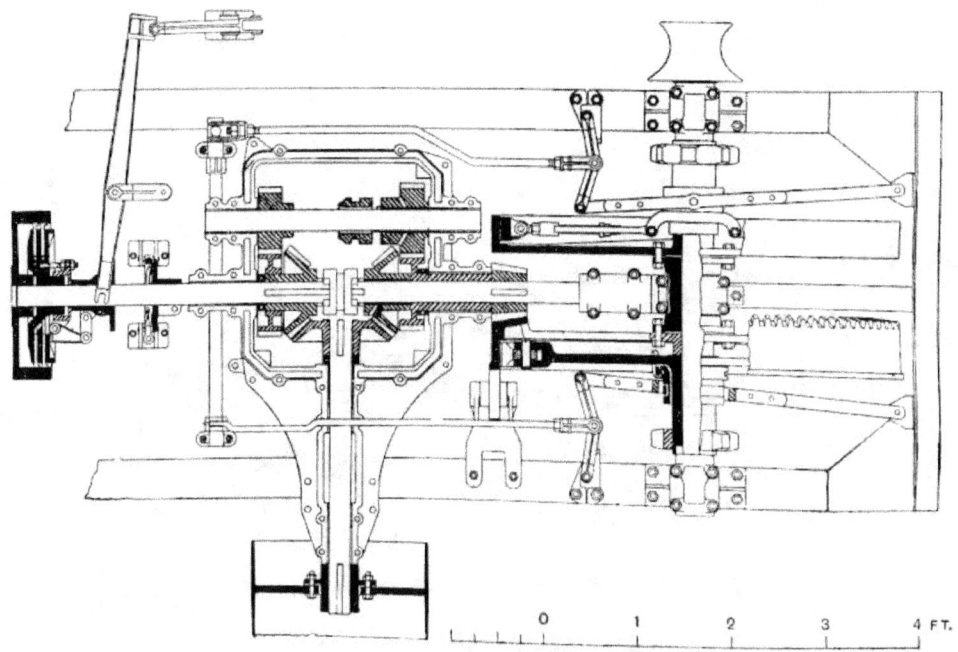

FIG. 87. – *Caterpillar; 75 h.p., Transmission, Plan.*

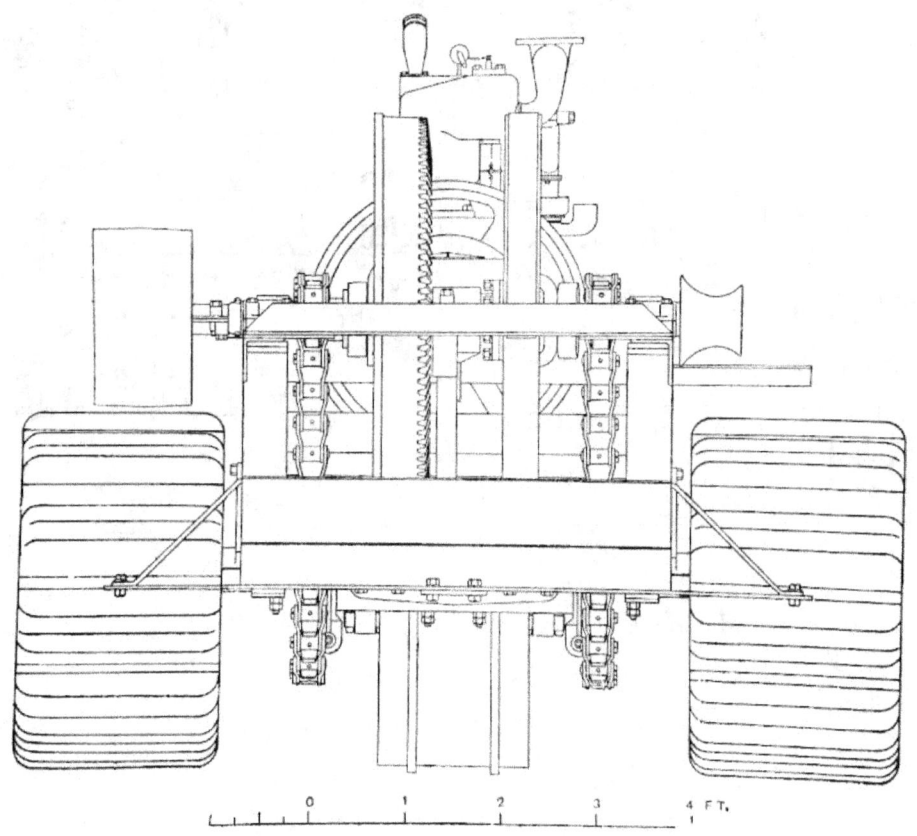

FIG. 88. – *Caterpillar; 75 h.p., End View.*

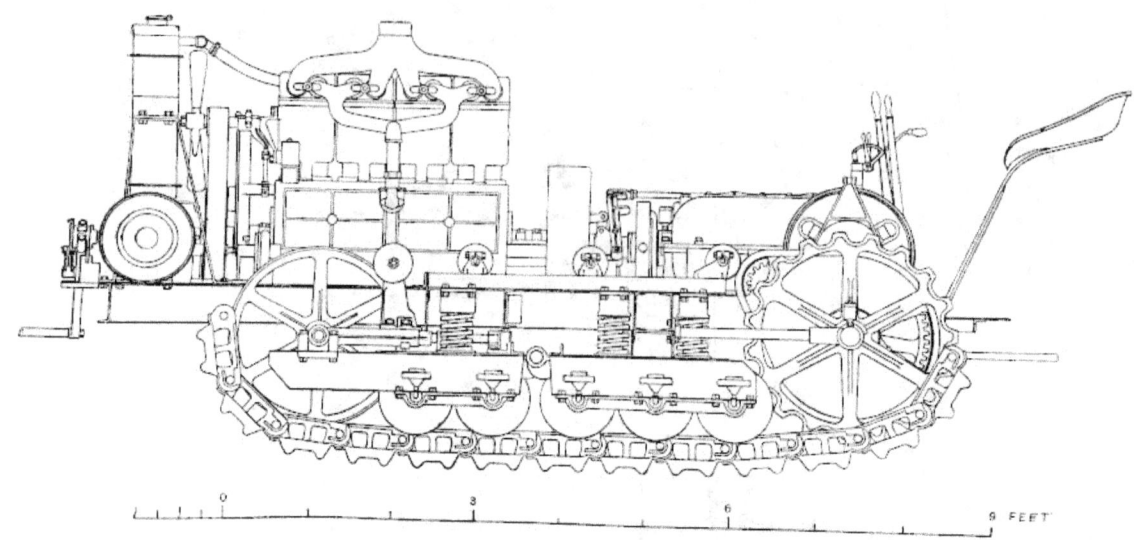

FIG. 89. – *Caterpillar; 45 h.p., Side Elevation.*

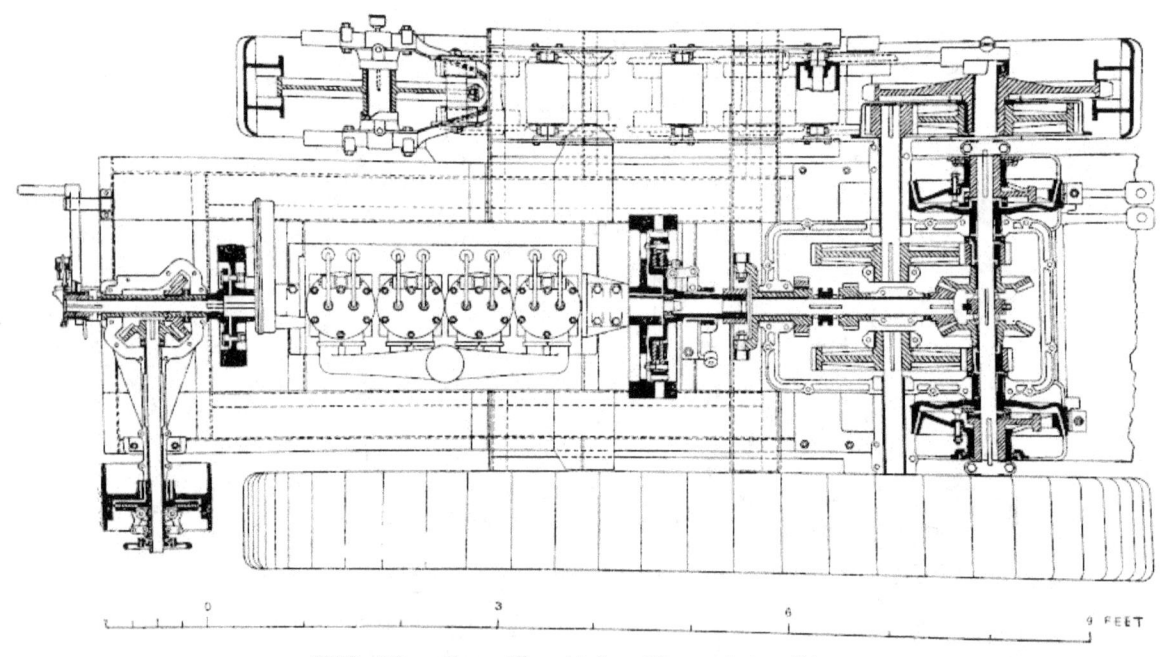

FIG. 90. - *Caterpillar; 45 h.p., Transmission, Plan.*

In the smaller models, 45 h.p., Fig. 76, Plate 14, and Figs. 89 - 90, and 18 h.p., Fig. 78, Plate 15, and Fig. 91, there is no front wheel, and. the steering is effected entirely by separate clutches and brakes ; the turning radius is controlled by pressure on a foot pedal. The truck-frame is itself articulated in these models, Fig. 51 (page 31), the leading half of the truck having two wheel axles and the back half three wheel axles, in both the 45 h.p. and 18 h.p. models; the springing of the back half of the truck from the main frame is very similar to that adopted in the larger models, but the front halves of the truck are connected by pin-joints to a transverse compensating beam pivoted horizontally and axially with the vehicle, Fig, 91, so that actually there is 3-point suspension and freedom for the truck-frames to accommodate themselves to irregularity of the ground.

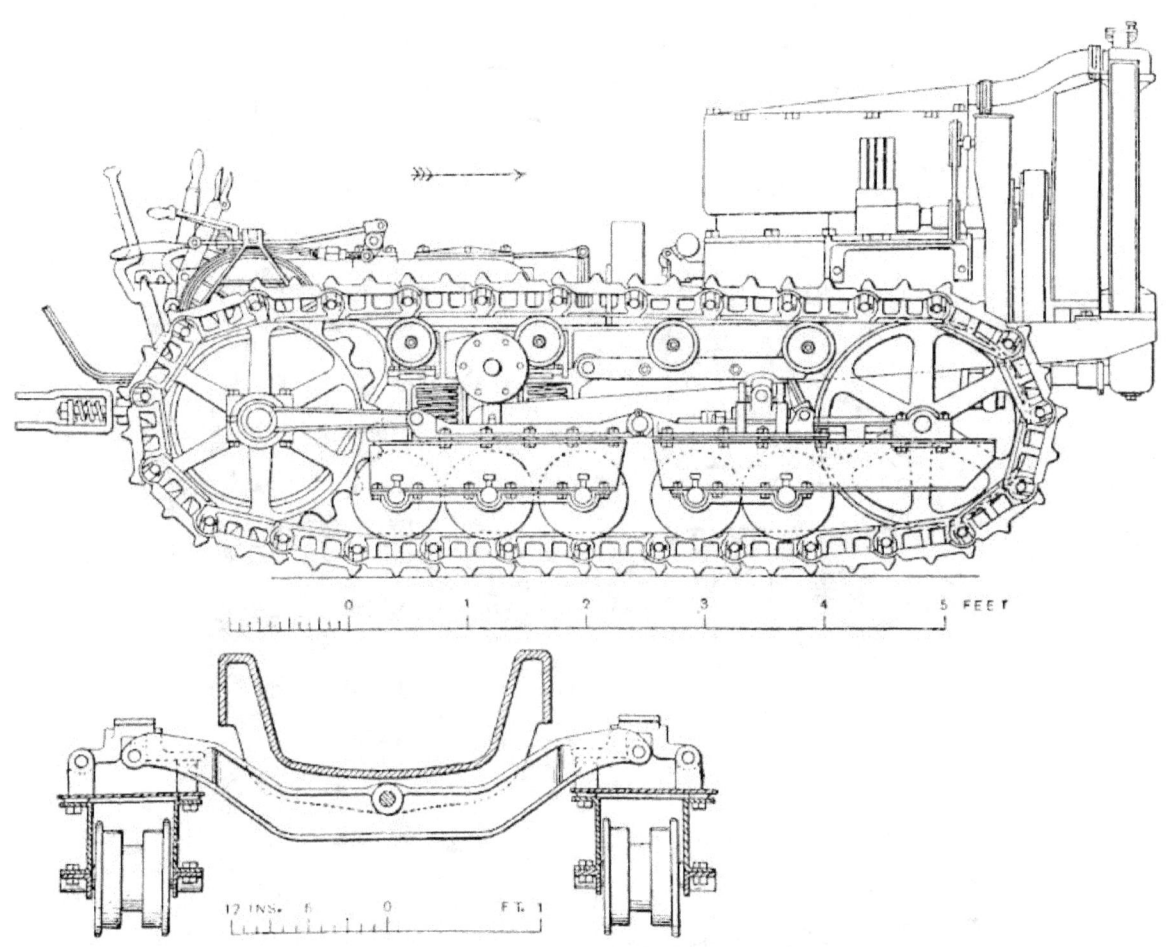

FIG. 91. – *Caterpillar; 18 h.p., Side Elevation and Detail of Three-point Suspension.*

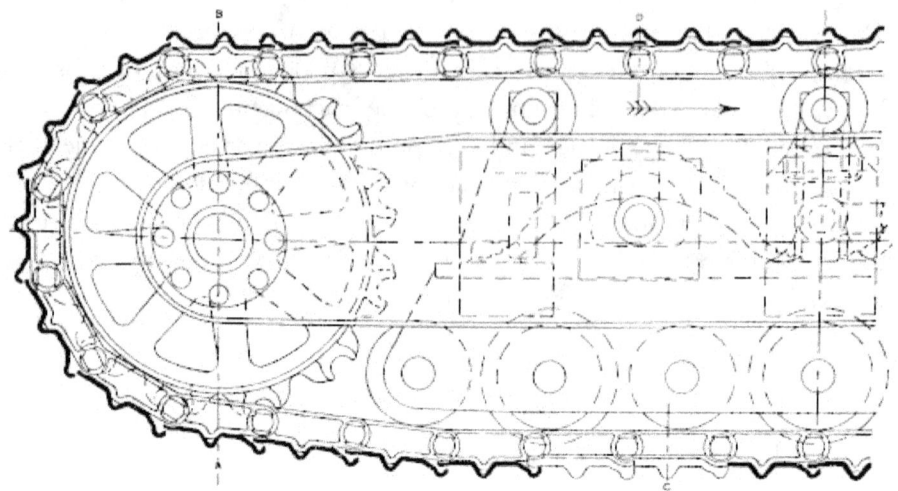

FIG. 95. – *Clayton 110 h.p. Tractor, Track-Frame and Details.*

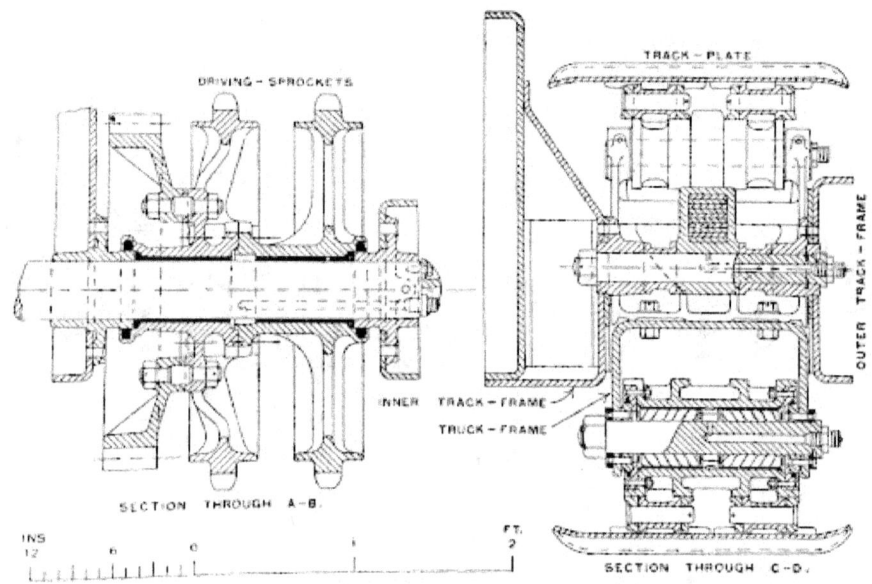

Detail of Chain-Track and Wheels; Enlarged View.

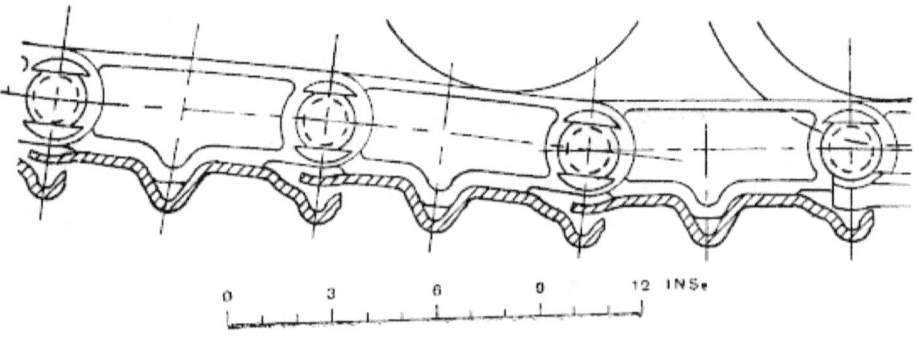

THE CLAYTON TRACTOR.

This tractor, Fig. 92, Plate 18, was produced by Messrs. Clayton & Shuttleworth jointly with Mr. W. F. Rainforth, Director of the Munitions Mechanical Transport Department. It is the sole representative of high-power chain-track vehicles of British design and manufacture, has fixed centres for the sprocket and idler as in the Caterpillar (Holt), but the arrangement of the truck is different, a single link connecting the truck to an anchor pin fixed to the track-frame, Fig. 48 (page 28). The load is carried on each truck by two inverted laminated springs about 2 feet long, each consisting of ten plates 3 inches by 0.3 inch, Fig. 95. There are seven supporting wheels, alternate wheels being flanged; those next the driving sprocket and the idler are without flanges. The length of chain unsupported between the end wheels and the sprocket or idler is nearly two chain-links in length. The links are built up of steel castings and pressed-steel shoes, Fig. 95; the latter have a central corrugation about 2 inches wide by 1.5 inch deep. Another corrugation is formed in the end of the sole plate that laps over the succeeding plate. The pin centres are kept very low, being only 2 inches from the under surface of the sole. An adjustment is provided for the bearings of the idler axle to take up wear.

The steering gear of this tractor is quite original in pattern, and consists of a single front wheel carried at the end of a projecting boom, Figs. 96 and 97. This is capable of being raised or depressed through a range of 15⁰ each way from normal, so that its position in relation to the track can be adapted to severe irregularities in the ground; it is not, however, necessary to use the elevating gear otherwise than for approximate adjustment, as a spring gear is fitted which provides for the range of movement in each direction.

Power is derived from a 110 h.p. petrol engine, constructed by the National Gas Engine Co.,* and is transmitted through a multiple-disk clutch with seven steel plates bearing on ferodo surfaces; the clutch shaft runs on ball-bearings. A flexible coupling with tripods connected by triple leather disks provides for possible errors of alinement between engine and gear-box, Fig. 98. The gear-box has three forward speeds and a reverse, change of gear being effected by means of the usual motor tractor sliding gear to a worm shaft, Fig. 99; the latter drives into a bronze worm-wheel, inside which the differential-gear pinions are carried. The differential bevels are each formed in one with a short hollow shaft, the interior of which is bushed to receive a short independent shaft, on which the worm-wheel centre is free to revolve. The ends of the bevel-wheel shafts are squared, and each fits into a driving pinion, which carries a brake drum on its inner side and engages with the gear-wheel attached to the twin sprockets. The sprockets are spigoted and bolted together, and to the spur wheel, and the whole arrangement is bushed and runs on the fixed back axle. All shafts are fitted with lubricating devices for enabling grease to be forced between the bearing surfaces. In each chain there are four bearing surfaces for the wheels, each 1.5 inch wide, giving an aggregate carrying width of 12 inches for the two tracks.

Steering is assisted by applying brakes independently to the pinion brake drums.

The Clayton tractor is shown in Fig. 93, Plate 18, hauling a test load, and in Fig. 94, Plate 19, climbing a 40⁰ gradient; the illustration shows how the front wheel leaves the ground at the top of the grade. The shock that would occur when the bank is climbed, and the machine again assumes the level, is in greater part or entirely absorbed by the springs controlling the projecting boom, Figs. 96 and 97.

* This motor was constructed to the designs of the Ministry of Munitions Mechanical Transport Department, the Director of which was Mr. W, F. Rainforth. A very full description of the engine is given in The Engineer, Vol. 123, pages 308, 310-11, 314 and Supplement.

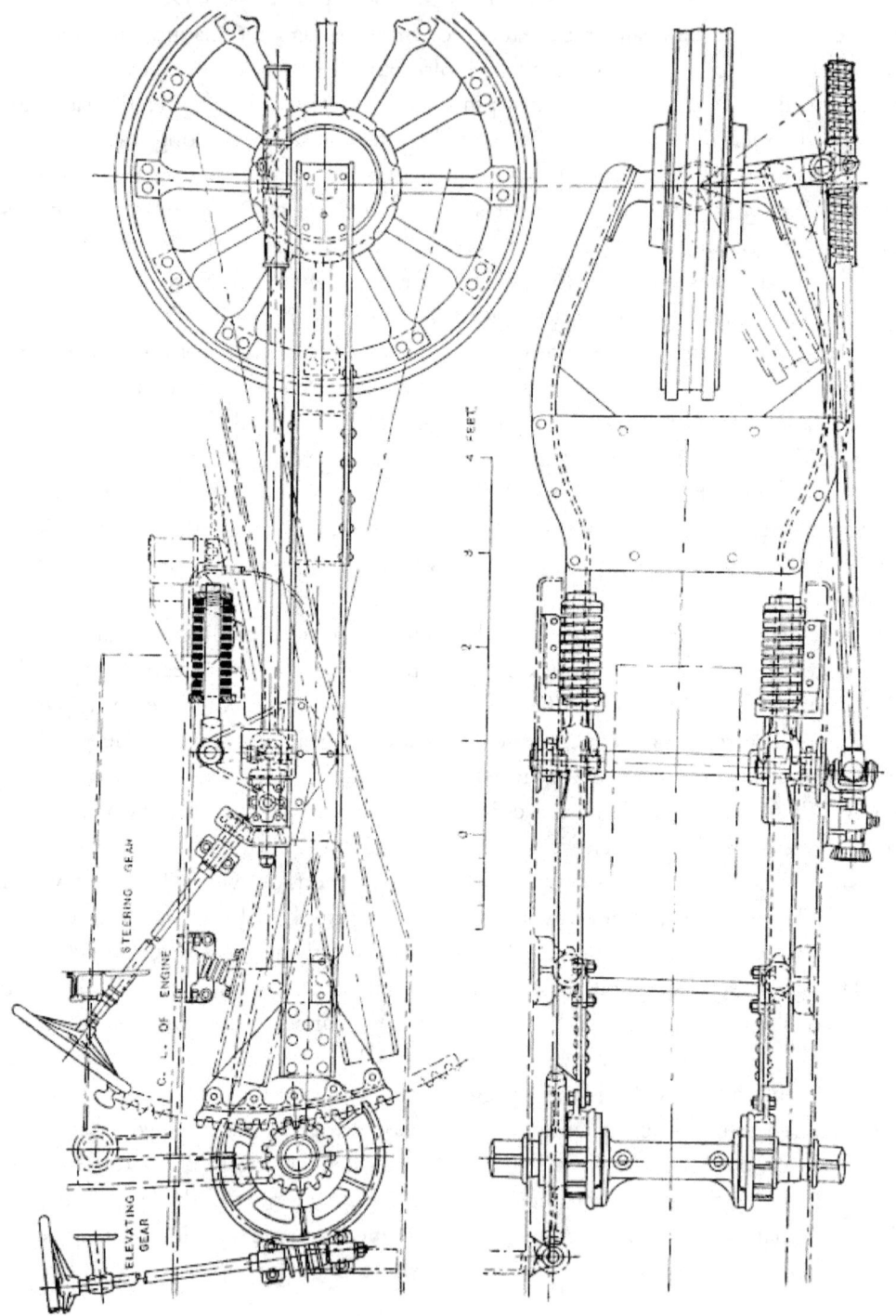

FIG. 96. *Clayton 110 h.p. Tractor; Front Wheel and Steering Gear.*

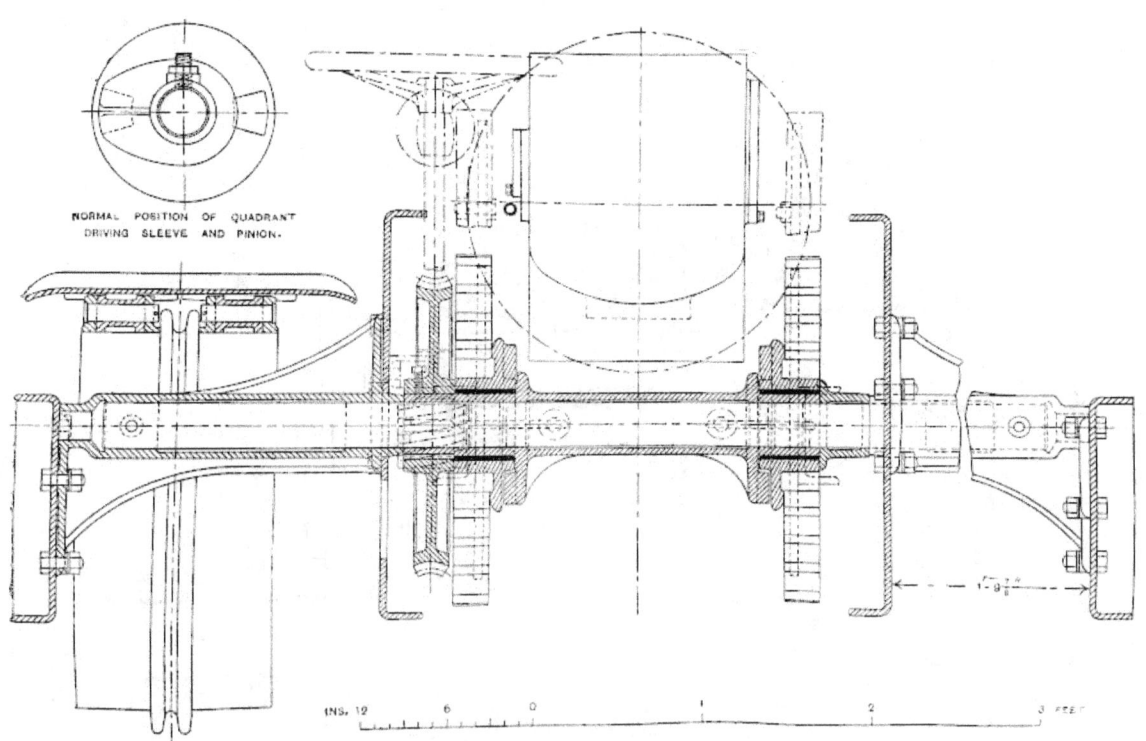

FIG. 97 - *Clayton 110 h.p. Tractor; Section through Fore-carriage.*

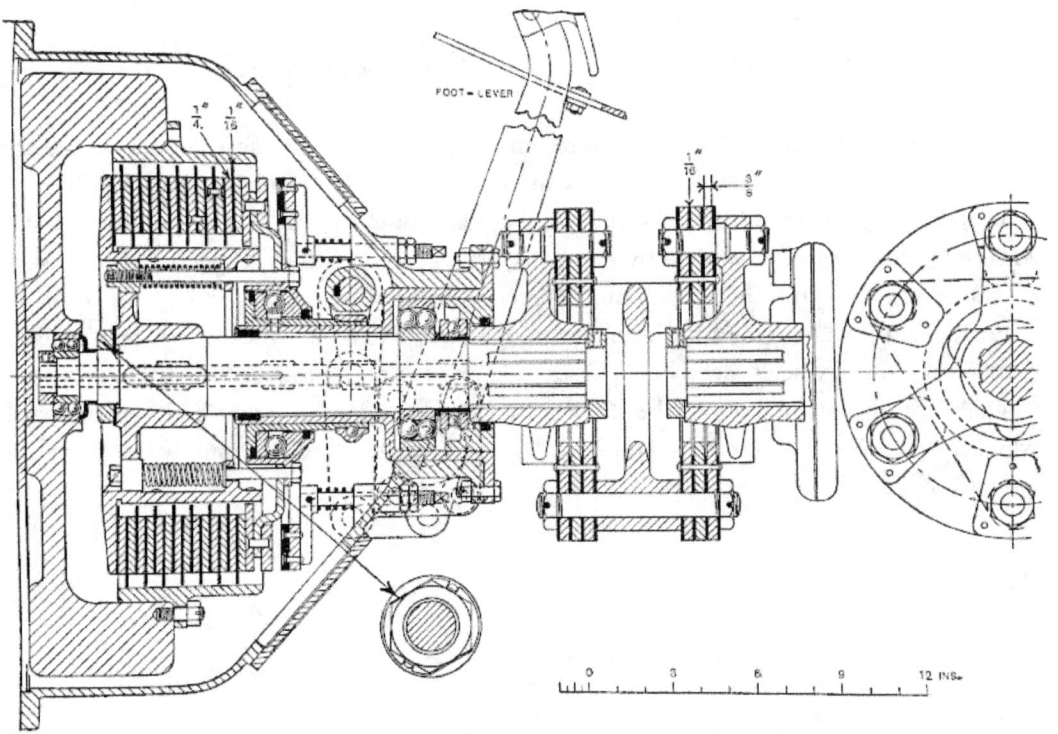

FIG. 98. - *Clayton 110 h.p. Tractor; Friction Clutch and Coupling.*

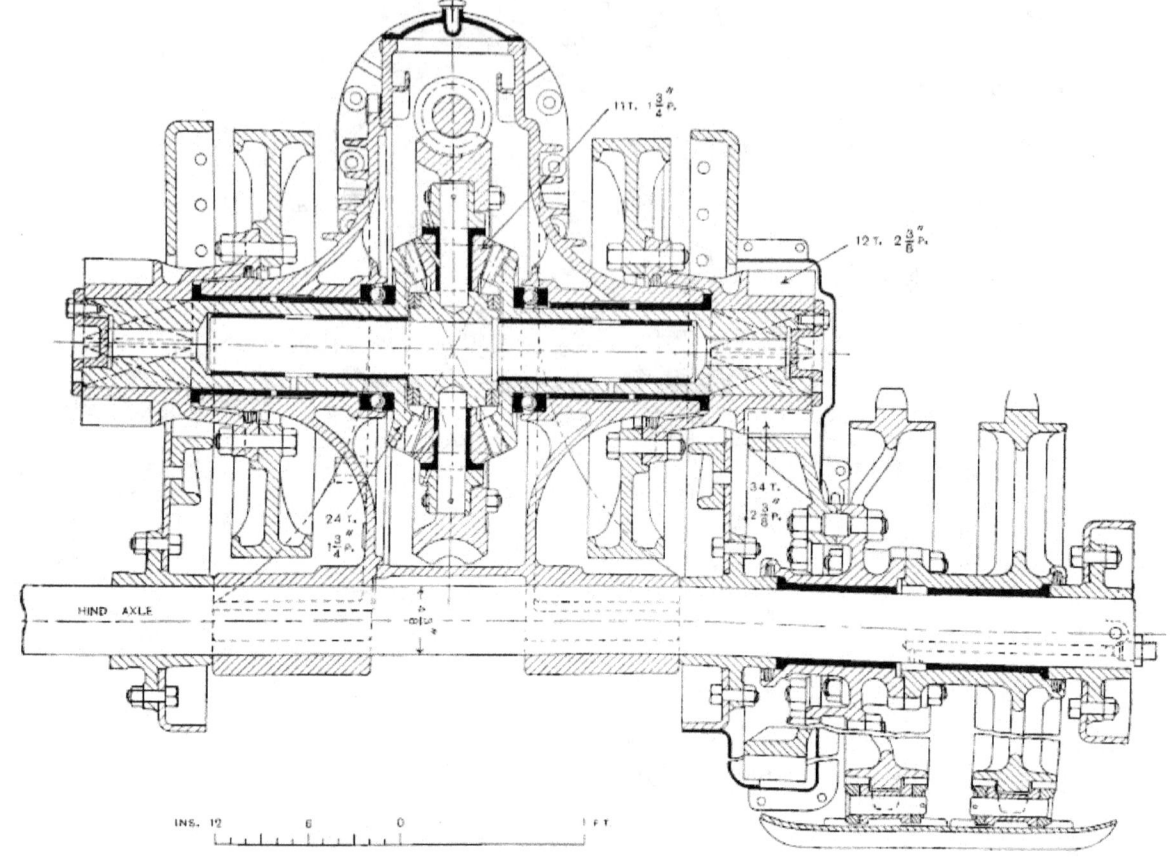

FIG. 99. – *Clayton 110 h.p. Tractor; Change-gear and Worm-box.*

The 35 h.p. Clayton tractor, Fig. 100, resembles the smaller American models in the absence of a front wheel; steering is effected by clutches and brakes operated through a steering wheel.

The change-gear provides two speeds forward and one reverse, and drives through clutches into two lantern pinion carriers on the driving-shafts, each gearing with its respective driving sprocket; the sprocket engages with the chain-track pins in every third tooth.

Each chain-track supports the vehicle through four fixed-axle wheels, the slack of the chain being carried on a single wheel; each chain passes over an idler at the front of the vehicle.

The tracks are pivoted about the driving shafts, the load at the front end being carried on a front axle which allows the two tracks to move independently of each other, and by means of a special connexion, to move only in a vertical path.

THE TRACKLAYER, C. L. BEST.

In both the 90 h.p., Fig. 101, Plate 19, and 75 h.p. Tracklayers the power is transmitted from the engine through an expanding clutch and flexible-spring coupling to the transmission gear-box, which is of the usual sliding-gear type. There are two forward speeds and reverse. The gears are cut from steel, and the differential gears are fitted with brakes for assisting steering; the steering gear can be actuated by a power-driven worm-and-wheel device or by hand. Each differential bevel and brake drum is integral with the pinion, which drives into an internal gear also cast in one piece with the driving sprocket, Fig. 103, Plate 20; the sprockets revolve on a fixed axle about which the track-frame is pivoted. The front of each track is thus free to rise and fall independently of the other, according to the irregularities of the ground, and the effect is that of 3-point suspension. The upper portion of the chain is carried on three or two wheels respectively, to limit sagging. The actual construction of the chain differs from that of other track chains that have been mentioned, as it is fitted with a rocker joint, Fig. 55, Plate 8, somewhat similar to that in use on the Westinghouse-Morse noiseless drive chain. This design is intended to work dry and to avoid the wear to which chains of the ordinary pin-joint construction are exposed when running in sand and grit. As in the other tractors described, an adjustment is provided at the front-end of the tracks for moving the idler-shaft bearings forward to take up wear.

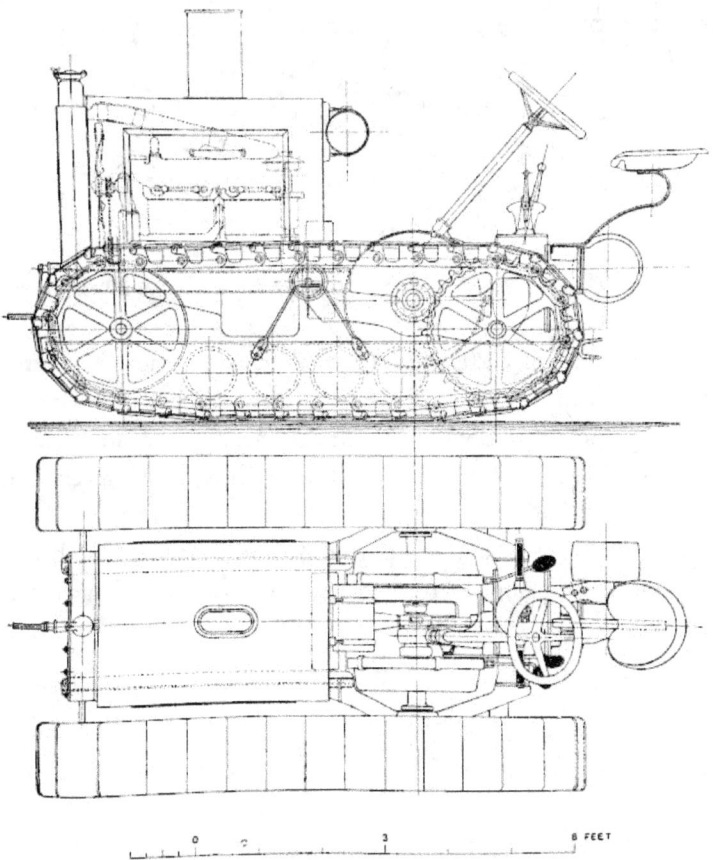

FIG. 100. – *Clayton 35 h.p. Tractor' Elevation and Plan.*

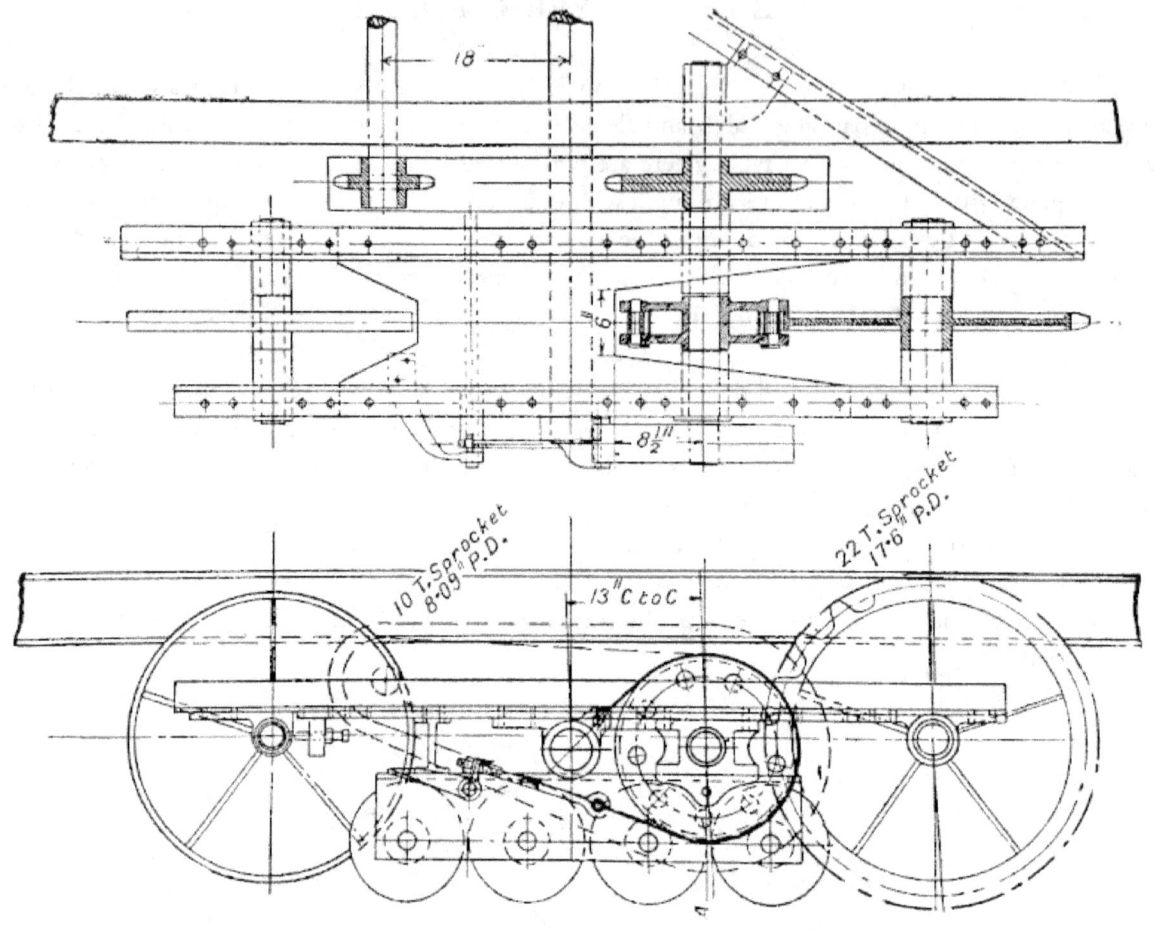

FIG. 105. - *Creeping-grip, Senior; 50 h.p.; Detail of Track-Frame and Drive.*

In the 30 h.p., Fig. 102, Plate 20, and 16 h.p. Tracklayers there is no front wheel, the steering being effected by side-clutches of the expanding type. Power is transmitted through a disk-clutch to the gear-box. In the 30 h.p. this is fitted with two forward gears and a single reverse gear; in the 16 h.p. there are two speeds forward and the same in reverse. In both types the track-frames are pivoted about the back axle and carried, near the front idler, by helical springs bearing on the main frame. In the 30 h.p. the load is carried on four wheels, and in the 16 h.p. on three wheels; in both these types the upper part of the chain is supported by a single carrying wheel. An example of the Tracklayer at work is given in Fig. 104, Plate 20.

THE CREEPING-GRIP TRACTOR, BULLOCK.

The chain-tracks in the Creeping-Grip tractors carry the tractor on wheels mounted on short axles fixed in the track-frame. The whole of each track-frame is pivoted about the axle, which passes through its centre, Figs. 105 and 106. This axle merely carries the track-frames. Power is transmitted from the gear-box to two pitch chains, which drive the two chain-sprockets; each of these is fixed to a short shaft running in bearings on the track-frame between the fixed axle and the sprocket; these shafts each carry a lantern wheel, which engages with the sprocket-wheel, the pins in the lantern wheel being of the same diameter as those in the chain-links, but the lantern wheel pins engage with each tooth of the sprocket, whereas the chain pins engage with the alternate teeth, Fig. 105.

In the 75 h.p. 'Giant,' and 50 h.p. 'Senior,' Fig. 107, Plate 21, models the weight is carried on four pairs of wheels in each truck; in the 30 h.p. 'Junior,' there are only three pairs of carrying wheels. The upper part of the chain is carried on three pairs of wheels in the two larger tractors and on one pair of wheels in the 30 h.p. model.

The gear-box is fitted with sliding dog-clutches, the gears being permanently in mesh; the differential gear is of the spur-wheel type. The motor of each model is governed. The front axle of the 75 h.p. and 50 h.p. models has Ackermann steering and a centrally pivoted leaf spring which bears on the two axle ends and supports the front of the tractor in the centre; the arrangement, which is similar to that of the Centiped (Phoenix), gives 3-point suspension.

The 30 h.p. model has a single front steering wheel carried in a forked frame and controlled by a worm and wheel gear.

The small 16 h.p. 'Baby, Fig. 108, Plate 21, has no front wheel, and is controlled by independent clutching of the two chain-tracks; it is possible to reverse either track independently for turning a circle of small radius. Details of the truck and of the track and wheels are shown in Figs. 110 and 111, Plate 22.

An example of the 16 h.p. Creeping-Grip at work is given in Fig. 109, Plate 21.

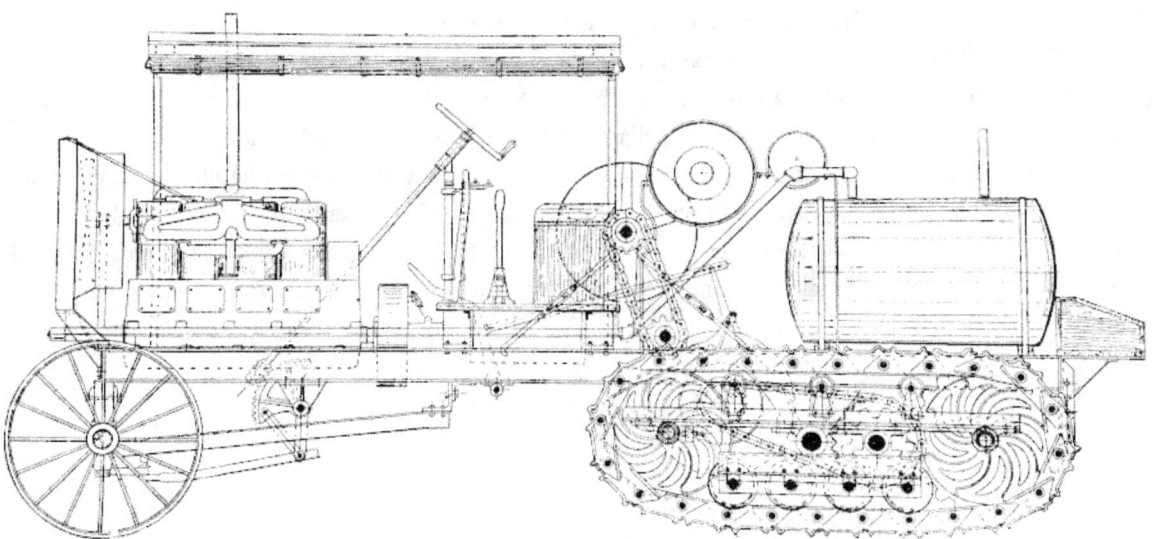

FIG. 106. - *Creeping-grip, Senior; 50 h.p.; Side Elevation.*

THE AUSTIN TRACTORS.

The F. C. Austin Drainage Excavator Co. of Chicago are makers of two models of chain track-tractors. The 35 h.p. tractor, Fig. 112, Plate 22, like most American low-power tractors is steered by declutching the tracks. This model is supported on six pairs of wheels to each track, which has a normal width of 12 inches, but the shoes, which are formed of flat plate with two corrugations, can be widened by bolting to them 36 - inch treads of wood reinforced with steel plate; the insistent weight can thus be reduced from 4.7 lb. per square inch to 1.78 lb. per square inch - the lowest figure given in Table 2. The arrangement obtained by bolting wide treads to the shoes is liable to the disadvantage of damage, or jamming by stones or sticks due to nut-cracker action; generally tractors working with very low insistent weight are used for cultivating swampy land on which stones do not exist.

The 15 h.p. Austin tractor, Fig. 113, Plate 23, is very similar in construction to the larger model described, but is supported on two pairs of wheels only for each track; the insistent weight is 4.7 lb. per square inch, and the tractor can be used for hauling a gang of four disk-ploughs, ditching, road making or grading.

BURFORD-CLEVELAND TRACTOR.

The Burford-Cleveland tractor, Fig. 114, Plate 23, is of much lighter construction than the machines previously described, having been designed with a view to the exigencies of work in this country, where it is necessary that the headlands should be kept as small as possible. The tractor has no front wheel, steering being effected by brakes on each side of the differential, operated by the steering wheel.

The load is carried by three pairs of wheels, and the slack part of the chain is carried on a single wheel.

The engine is of the ordinary four-cylinder type carried in the main-frame by three-point suspension. The exhaust is led through a sleeve jacket on the lower part of the induction-pipe; this allows the use of inferior fuel such as paraffin or creosote waste.

The front end of each track-frame is formed with a yoke encircling the track-chain, and is connected by links to the end of a cross-spring the centre of which is secured to the main-frame by U-bolts in the usual manner; the tracks can oscillate about the axis of the back-axle.

The Burford-Cleveland tractor is shown in Fig. 115, Plate 24, at work hauling two ploughs.

It is to be noted that in all the vehicles described of this class the torque applied to the sprockets, which tends to lift the front of the vehicle, has the further effect in cases where the track-frame is pivoted, as in the Tracklayer and Creeping-Grip tractor, of increasing the insistent load on the front of the track itself.

THE STRAIT TRACTOR, KILLEN-STRAIT.

This tractor, Fig. 119, Plate 26, presents several original features; the unsymmetrical driving track-frame, with its axle placed high and far back, carrying the driving sprocket over which the chain passes; the series of three pairs of large-diameter wheels under which the chain-track runs; and the spring arm which, by means of a helical tension spring, keeps the front of the track in contact with the ground.

The drawbar is attached to the frame by a swivel joint forward of the back axle, and it swings clear below the axle. The motor drives through a cone-clutch to the gear-box, which is of the sliding-gear type with reverse through bevel gears; the drive to the differential, which is of the bevel-gear type, is effected through spur gears; one driving sprocket is carried on a sleeve to which one bevel wheel is fixed, the other bevel wheel and sprocket being both keyed to the axle. The steering is operated by worm and wheel fixed to a track-frame supported on a helical spring. This frame carries a miniature track of fifteen treads, Fig. 116, Plate 24. The load is carried through three pairs of wheels to the track-frame, Fig. 118, Plate 25; the front axle of these is adjustable for taking up wear. In climbing steep banks or in crossing ditches this front track is sometimes quite off the ground, behaving somewhat similarly to the leading wheel of the Clayton tractor, but with the difference that, whereas in the Clayton tractor the track-frame is fixed and the whole movement taken by the front wheel, the movement in the Strait tractor is partially taken up by the spring on the front track column and partially by the springs that control the rocking action of the driving-tracks.

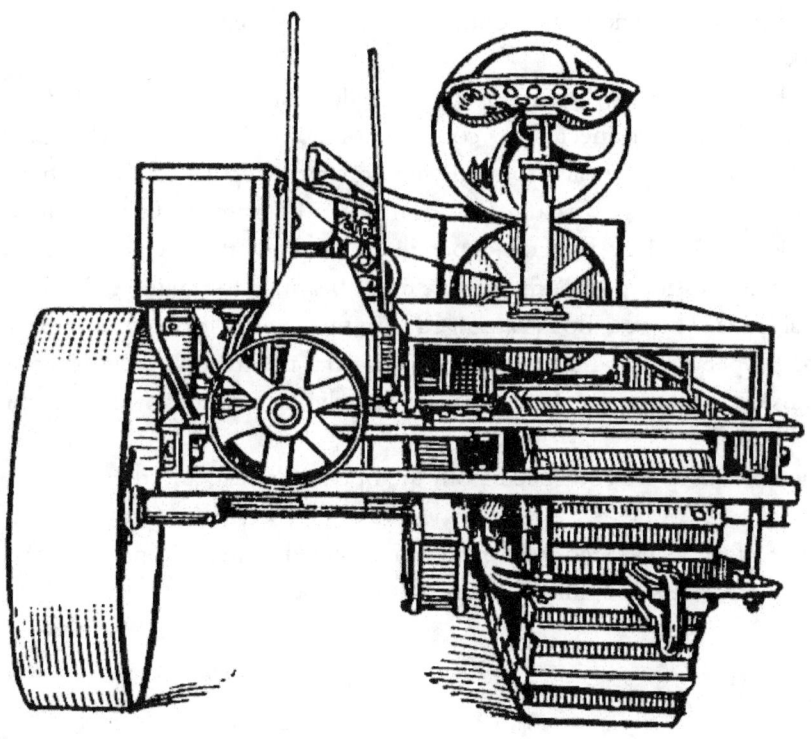

FIG. 123. - *Strait Tractor; Model 3; End View.*

In the Strait model 3 tractor, Fig. 122, Plate 27, the power is applied to a single chain-track at one side of the vehicle; this track is of similar construction to that of the ordinary model just described. The steering track carried in front is also similar to that already described both in form and in spring suspension, but it travels in the same path as the driving track. The engine is carried to one side of the main or driving track, and the machine is supported on the extreme left by a wide carrying wheel or roller, which runs freely, but can be adjusted laterally on its spring-supported axle, Fig. 123.

The gear-box is fitted with a single gear and reverse, but the engine speed is made variable, as in the ordinary motor lorry.

Examples of work performed by the Strait tractor are given in Figs. 120 and 121, Plates 26 and 27, and the model 3 is shown in Fig. 117, Plate 25 on deep ploughing.

The arrangement of the draw-bar attachment in the Strait tractors is kept low and forward so as to increase the load on the front of the tractor when hauling and consequently to gain adhesion.

THE BALL-TREAD TRACTOR, YUBA.

In this form of tractor, Fig. 124, Plate 28, the weight is distributed over the chain-track by two rows of steel balls 2.25 inches diameter, and is carried on ball races of annealed cast steel, built into the track-frames. The ball-race sections carried on the shoes are of L-form; those on the track-frame are, in section, of the form of a circular arc. Thus each ball bears against two faces at right angles to each other on the shoe and against the appropriate portion of the curved section of the track race, Figs. 127 and 128, Plate 29; it is claimed that this method of carrying the side-thrust renders the arrangement peculiarly suitable for use on side-lying ground. The ball races are kept supplied with heavy lubricating oil when at work.

The track-frames are pivoted about the rear axle, which also carries the driving sprockets; the front of each track-frame is connected, by compression springs, with a bracket carried on the main-frame. The driving sprockets are each formed with an internal spur-gear, into which gears the pinion carried on the corresponding cross-shaft of the gear-box. The clutch gear is such that the drive to each track, Fig. 129, Plate 29, can be reversed independently, enabling the machine to be turned in about its own length. The transmission gear is fitted with roller bearings for taking radial loads, and with ball-bearings for axial loads. The gear-box is arranged with two speeds forward, and the same for reverse.

Steering is operated by wire-rope connexions from the steering-wheel shaft to the front wheel frame, which runs on ball-bearings. An example of the climbing power of this tractor is shown in Fig. 132, Plate 31.

The speed changes are controlled from the operator's seat, and there is a foot accelerator for the motor.

Examples of the Ball-Tread tractor travelling on side-lying ground and at work are shown in Figs. 126, 130, 131 and 132 - 136, Plates 28 and 30-32.

MARTIN'S AGRICULTURAL TRACTOR.

This is a vehicle of lower power than the majority of those described above; it has been designed with the object of meeting British rather than Overseas conditions. Each track-chain passes round two idlers on the track-frame and over the track-sprocket carried on the main cross-shaft, Fig. 137, Plate 33. Either sprocket wheel or both can be engaged with the cross-shaft by means of dog-clutches. The tracks are arranged at the front of the machine, which weighs about 30 cwt. including the three-furrow plough.

Steering can be effected by a single central wheel, carried in a castor frame, for use when ploughing, or by a pair of steering wheels mounted in a carriage and operated by means of a pinion and quadrant, Fig. 138, Plate 33.

Examples of the Martin tractor at work are given in Figs. 139 to 141, Plates 33 and 34.

THE WOLSELEY MOTOR-SLEIGH.

This vehicle, Fig. 142 (page 61) and Fig. 143 to 146, Plates 34 and 35, was designed and built for Captain Scott's Antarctic expedition in 1909; three of these motor-sleighs were made with the object of facilitating transport over snow and ice.

A previous attempt was made by Mr. B. Hamilton to design a vehicle to cross hummocky ice and lumpy snow and ice, but on test in the Alps it was not found to meet the conditions sufficiently well. The Wolseley machine was designed to work with chain-tracks and to form a motor-sleigh. The first vehicle, which was completed in February 1910, passed its tests satisfactorily on Lake Fefor in Norway, Fig. 146, Plate 35. The vehicles were used by Captain Scott, and rendered much assistance in the earlier stages of the expedition, but were left behind in the final dash for the South Pole.

The vehicle is remarkable in the arrangement of bearers which take the load; these are of wood, shod with sheet aluminium to avoid 'snow-clogging.' The chains were of soft steel links with wooden rollers. It is to be noted that the arrangement adopted differs from any of the other tractors described, as the rollers travel with and at the same speed as the chain, and the rollers also form the gearing surface with which the driving and idler sprockets engage. The heads of the bolts that secured the links to the chain-feet were made of pointed form, as shown in Fig. 142, so that the vehicle could travel equally well on smooth ice or over snow so loose that a man could only travel over it on skis. The engine was of 12 b.h.p., air-cooled, transmitting its power through a two-speed box to a differential-box and axle at the back end. The front axle carried two idlers. The speeds provided were 1.75 mile per hour on low gear and 3.5 miles per hour on top gear. In the trials made at the works it was found possible to drive the machine over loose ashes without scattering them in any way. The motor-sleigh was capable of climbing with ease a gradient of 1 in 2 (about 27°).

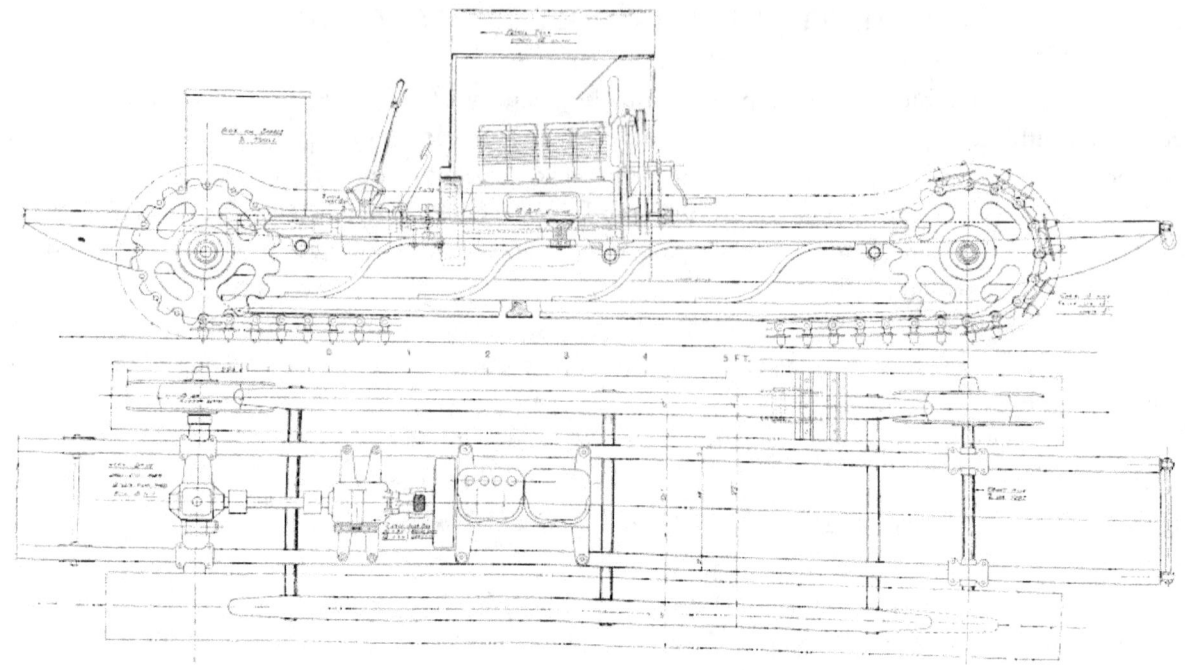

FIG. 142. – *Scott Motor-Sleigh*.

MIXED TRACTORS.

By mixed tractors are meant tractors in which the traction can be effected wholly or in part by chain-tracks, the load being carried on wheels; the only example of this class known to the Author is the Lefebvre tractor.

THE LEFEBVRE TRACTOR.

This vehicle, Fig. 147, Plate 35, is designed to travel on wheels when running over the ordinary road, the tracks being raised out of contact with the ground.

The tracks are of a construction quite different from any of the heavy-weight carrying tracks previously described. In this instance light steel casehardened spade-plates of L-section are attached to the pitch chains, each plate bridging 6 links of the chain. The spades present a vertical face to the ground, into which they cut when the chain is assisting the wheels, The aggregate surface so presented is that of 7 plates on each track measuring about 4 inches by 6.5 inches, giving a total area of about 360 square inches, available for propulsion on soft ground on which the wheels would not grip. These chains on each track are carried on a frame, pivoted about the rear or driving sprocket, which is not concentric with the axle of the back wheels, but placed at some distance behind the axle. A jockey pulley carried on the track-frame depresses the chain slightly in advance of the rear sprocket. Vertical guides with a power-operated screw-gear enable the track-frames to be depressed or raised through the full range in fifteen seconds under control of the driver.

The adherence chains are so geared as to travel at the same linear speed as the periphery of the driving wheels, but a dog-clutch carried on the sprocket shaft enables the chain-drive to be disconnected when the chains are elevated and the tractor is running on hard ground.*

It is claimed that with the assistance of the adherence chains the tractive effort can be as much as 50 per cent of the weight of the vehicle.

* A description of this tractor is given in *Engineering*, Vol. 101, pages 252 - 3.

APPLICATIONS OF CHAIN-TRACK TRACTION.

The following are examples of work done by chain-track tractors where roads are poor or absent, and heavy loads require to be moved:-

Timber. - 80,000 feet of green rough timber hauled 15 miles over snow and ice track at a temperature of - 40° C.; the return journey of 15 miles without load but with empty sleighs. Water tank and caboose for crew are included in the load both ways.
Clay and five to six wagons, loaded each with 6,000 lb. over heavy sand and over black land. Ploughing virgin country; turning a 30-inch furrow 12 inches deep through scrub, Fig. 71, Plate 13.
Ploughing with five 14-inch ploughs to 8 inches depth, or with six 14-inch ploughs to 6 inches depth (45 h.p.).
Ploughing 40 acres per day (75 h.p.).
Ploughing partly drained marsh land.
Ploughing volcanic ash soil.
Ploughing a hill side.
Ploughing 16 inches deep in Texas Panhandle.
Ploughing 12 inches deep in black adobe.
Ploughing with a gang of 21 disk-ploughs.
Ploughing in sugar plantations, Cuba.
Ploughing 15 acres per day (35 h.p.).
Ploughing 100 inches wide by 7 inches deep in sandy loam (35 h.p.).
Rice farming; ploughing, harrowing and seeding. The chain-tracks enable irrigation ditches to be crossed; the tractors also haul harvesters and binders in the harvest season and do the work of ditching.
Ditching; hauling a ditcher for laying tile ditches, Fig. 82, Plate 16.
Road making and grading.
Hauling materials through desert sand for Los Angeles Aqueduct.
Clearing land for Catskill Aqueduct.
Hauling rock, ore, and road material.
Hauling load of 17 tons up a 20 per cent grade in Utah (75 hope).
Hauling load of 40 tons of rock across country for highway construction in California (75 h.p.).
Hauling piledriver in marshy land.
Hauling house weighing 50 tons 1b mile in forty-five minutes on dirt road, Fig. 120, Plate 26 (50 h.p.).
Hauling excavator.
Hauling scarifier.

Hauling sprayers.

Hauling harvester for 20-feet cut.

Hauling 8-feet double disk cultivator.

Hauling and operating hard - pan driller on fig plantation, California, for planting by dynamite, Fig. 133, Plate 31.

For all the above operations it is estimated that the cost of haulage by tractor is from one-third to one-half of that by horse or mule.

In the instances quoted a few only of the appliances hauled or operated are designed specially for use with chain-track tractors.

CHAIN-TRACK HAULAGE WAGONS.

Where heavy loads have to be carried over sandy and boggy country or through wide shallow fords with soft bottom, it is possible to extend the principle to haulage wagons fitted with chain tracks; by this means a 100 h.p. tractor can haul a train of four wagons, weighing 5 tons each, and carrying a load of 10 tons each, or a total dead load of 60 tons and a paying load of 40 tons, over soft ground capable of supporting only 7 or 8 lb. per square inch.

Tractors hauling trains of chain-track wagons are shown in Figs. 148 and 149, Plate 36.

The Pedrail, fitted with an improved roller chain, Fig. 54 (page 37), arranged with two sets of rollers with axes at right angles to take vertical and horizontal. loads respectively has been tried on a haulage wagon for stone, Figs. 150 and 151, Plate 36. The rolling friction is reduced by this arrangement to so small an amount that the vehicle will roll back on a 2 per cent slope.

OTHER APPLICATIONS OF THE CHAIN-TRACK.

Trenching, ditching and farm drainage machines with chains of dredger buckets or excavating wheels are in many instances carried on and propelled by chain-track supports.*

* Since this Paper was written a farm-drainage machine, with excavating wheel and chain-tracks, constructed by the Pawling and Harnischfeger Co. of Milwaukee, Wis., has been described and illustrated in *Engineering*, Vol. 104, pp. 228 and 237 - 8.

PARSONS EXCAVATORS.

The Parsons Company, of Newton, Iowa, make nine models of trench excavators; particulars of six of the larger models are given in Table 2 (pages 78 - 83).

The model 60 Trench Excavator, Fig. 152, Plate 37, is carried as to one-half of its weight on chain-tracks, the other half being distributed over pairs of leading and trailing wheels. The machine is steam-driven and comprises a conveyor for landing the spoil at an adequate distance from the trench.

The model 48 Trench Excavator is carried as to 75 per cent of its weight on chain-tracks, the remainder being carried on the leading wheels; the machine is driven by an internal-combustion motor of 62 h.p., and is illustrated in front-view in Fig. 153, Plate 37.

The model 24 Trench Excavator is of smaller size and fitted with a 30 h.p. motor; 71 per cent of its weight is carried on the chain-tracks and the remainder on the leading wheels. It is shown in back view in Fig. 154, Plate 38.

Of the other models one is entirely carried on chain-tracks, the steering being effected by clutches as in the smaller models of chain-track tractors (Burford-Cleveland, etc.).

AUSTIN DRAINAGE EXCAVATORS.

The F. C. Austin Drainage Excavator Company are makers of seven models of trenching or ditching machines capable of digging trenches from 6 feet to 25 feet deep, and from 15 inches to 72 inches wide. The rate of digging is stated to vary from 10 feet to 3 inches per minute.

The motors are from 18 h.p, to 85 h.p.; one of these machines is shown in Fig. 155, Plate 38.

The spoil can be delivered either to the right or left or to both sides of the machine as may be desired.

Another series of four models is constructed by the Austin Company and. arranged for performing the further operation of bank-sloping for the sides of the trench either as a single operation, or by roughing and finishing operations. The wheel of the excavator, over which the chain of buckets passes, is formed of two truncated cones, arranged base-to-base, the conical surface being represented by sloping cutter-bars capable of cutting away hard adobe soil. The wheels are constructed for various slopes of bank from to 1, to 1 to 1; the machines can be used for preliminary work in railroad construction, trenches being cut on each side of the land to be used, and the spoil from the two trenches conveyed to the middle of the strip where it is made up to formation level, bermes of the desired width being left on each side between the ditch and formation as shown in Fig. 156, Plate 39.

Those ditching machines which are intended to perform consecutive operations of trenching and bank-sloping have the chain-tracks arranged to straddle the trench; this necessitates great breadth of the machine which will measure as much as 14 feet 6 inches between the insides of the chain-tracks for cutting a ditch 10 feet wide at the top.

APPLICATIONS OF TRENCH EXCAVATORS.

These comprise: cutting trenches for sewers, water-mains, gas-mains, electric light and power cables, as well as work for street railways and railroad construction.

BACK FILLERS.

The land dredgers having cut the ditches or trenches at high speed to the required depth, and the sewers or mains having been laid, it is necessary to fill in the spoil again. For this purpose a different class of machine is used, which is virtually a locomotive crane also carried on chain-tracks. These machines are made by the constructors of trench-excavating machinery, and handle the spoil either with a clam-shell or grab bucket or with a drag-line bucket or a combination of the two for use alternatively. An example of a back-filler is shown in Fig. 157, Plate 39.

Further extensions of the use of the chain-track for carrying pile-driving machinery, as well as that used for road-making and for deepening channels in shallow streams may be expected in the future.

In conclusion it can be safely said that where a track can be found seven feet wide which does not present rocky obstacles over ten inches high, whether over sand, ash, clay, marsh, snow or ice, on the level, or on gradients up to 30 per cent, the chain-track tractor has proved itself capable of travelling and performing useful work beyond the capacity of animal haulage.

For this reason the chain-track tractor may be expected to occupy a position of ever increasing importance in the development of new countries, and in places where it is necessary to transport machinery and stores over roadless country many miles from the railway.

The difficulties presented by the working conditions are such that the Author cannot regard any one of the existing systems as perfect, and expects that much will be done to improve the vehicles in the near future. There is probably room for considerable improvement in the chain-track itself as well as in its supports and its anti-friction devices. The variation in resistance to haulage is so great as between one system and another, not merely in static resistance to starting, but in dynamic resistance to haulage, that, as already suggested, the various track systems should form the subject of comparative tests at some of the great agricultural colleges.

One feature that is particularly remarkable is the small amount of power lost in compressing soft marshy land under the feet of the track-chains. A very visible alteration of the appearance of a marshy surface due to the prints made by the track feet may, for example, only involve an expenditure of power which is but slightly greater than that required for hard ground, whereas the effort involved in walking over the same soft surface may be many times greater than that of walking on the hard road.

The Author desires, in conclusion, to express his thanks to Dr. William H. Maw, Past-President, I.Mech.E., to Mr. Coker F. Clarkson, General Manager of the Society of Automobile Engineers, to the Editor of *The Engineer*, to Col. R. E. B. Crompton, C.B., R.E., Dr. H. S. Hele-Shaw, and Messrs. W. Defries, W. F, Rainforth, L. St. L. Pendred and G. W. Watson, MM. I.Mech.E., as well as to the Engineers, representatives, and agents of the firms manufacturing the vehicles described, for information on a subject which is at the present time very deficient in literature, for valuable assistance, and for very many of the photographs and drawings from which the illustrations for the Paper have been prepared.

The Paper is illustrated by 40 Plates, 50 Figs. in the letterpress, and is accompanied by 2 Appendixes with 4 Figs. and 2 Tables.

APPENDIX I.

APPENDIX I.
HISTORICAL NOTE.

The use of multiple-wheel traction for road-tractors appears to have made its appearance in practical form almost simultaneously in France and Holland about 1903; in the latter country as the Spyker passenger car, and in the former as the Renard road-train. The idea of the road-train was taken up by the Daimler Co. of Coventry in 1907, and trains on this principle, embodying considerable improvements on the original French design, were tested in India. Apart from trials on comparatively flat country near Bombay, other and very severe trials were made on the Gauhati-Shillong road at the instance of the Indian State Railway. This road, which is about 64 miles in total length, with about seven miles level and a rise of nearly 5,000 feet in the remainder accompanied by the sharp curves and steep pitches of an ordinary cart road, was subsequently worked on the block system with single vehicles. The abandonment of the Renard train does not, however, appear to have been due to any failure to meet the conditions of traction.

An illustration of the complete Renard road-train is shown in Fig. 158, Plate 40, and the chassis are shown connected in plan in Fig. 159, Plate 40.

The vehicles were connected for traction by a coupling - bar which ensured the steering of the trailing vehicles; the power was transmitted from the locomotor by a cardan shaft, with a sliding joint in the coupling-section to permit of relative rise and fall of the vehicles.

There does not appear to have been either provision for compensating for varying velocities of the wheels of different vehicles, which could occur even with wheels of equal diameter on uneven surfaces, or provision of automatic locking devices for the differentials in case any driving wheel failed to grip. In such event the drive was taken by the other vehicles, but neither of the driving wheels of the vehicle in question could assist in traction under such circumstances.

It is worth noting that, in Young's work,* published in 1860, it is mentioned that a quarter-scale model made by Mr. Burstall of London in 1827, and tested at Edinburgh where it ran 250 miles in eight days, was arranged with a differential gear similar to that fitted to a steam coach by Rufus Porter of Hartford, Conn., about 1830. The date of Burstall's model is therefore earlier than the date of Roberts's invention previously given by the Author in a Paper on the "Development of Road Locomotion in Recent Years."† According to Young, the mitre gear arrangement of differential also appears in the subsequent American carriage of James about 1834. From this it would seem that the idea (if the application of the balanced differential gear to traction originated in Britain, but was first applied to full size machines in the United States.

Difficulties with the bevel or spur forms of differential drive are familiar to all car drivers who have had experience with chain drives, particularly the difficulty caused by the fracture of a chain, which prevents driving except by the alternative of locking one sprocket and driving the other through the differential gear, with resultant doubling of the speed of the corresponding road wheel.

---

* The Economy of Steam Power on Common Roads, page 351.
† Proceedings, I.Mech.E., 1910, page 1555.

To overcome this difficulty Hedgeland, of America, produced a solid axle with cone-clutches inside the wheel hubs, internally threaded, and carried on screwed portions of each end of the back axle, Fig. 160. The locking of one or other of these clutches with the corresponding wheel-hub permitted driving; overrunning of the wheel unscrewed the clutch and freed the wheel. On curves the work was consequently all thrown on to the inner wheel, or, if the wheels were of unequal diameter, on the straight the larger wheel alone became the driver; furthermore, if the brakes were applied or other change made in the drive, causing engagement of either or both of the clutches, appreciable shock resulted.

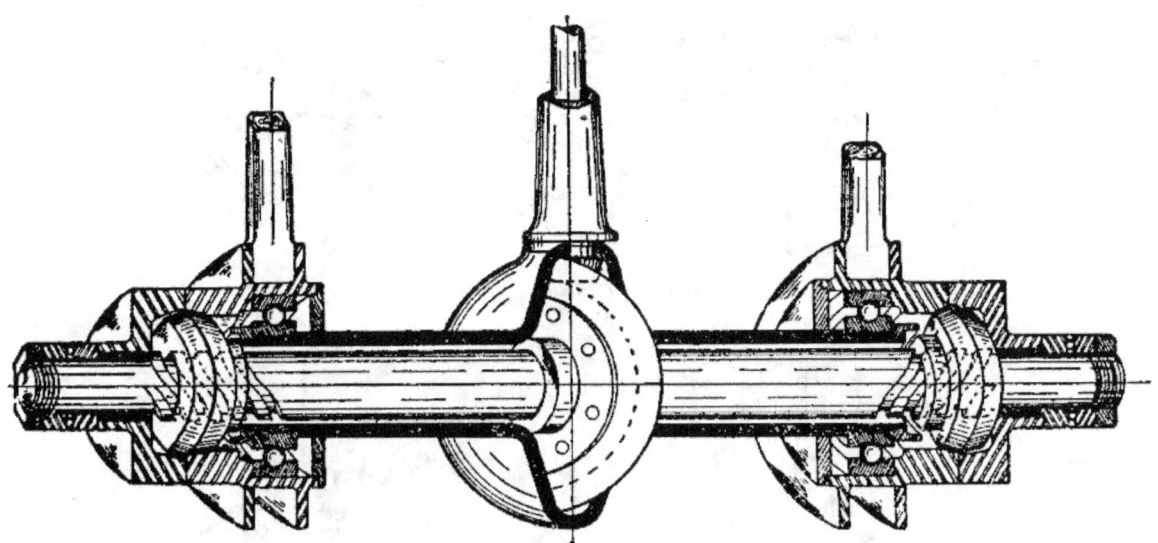

FIG. 160.- *Hedgeland Axle.*

It will be seen that with the Hedgeland axle the car would be driven by either back wheel on the straight at a speed corresponding to the normal mean revolutions which would be obtained with the ordinary type of differential, but that on a curve it would be driven faster than the normal.

This class of differential is therefore not a true differential, in the sense that it does not divide the power; but it is an arrangement equivalent to free-wheeling either road wheel in advance of the propelling gear. Several forms of cam and roller (derivatives of the Bourdon clutch) or eccentric gears have been devised for the same object, all of which are merely mechanical variants of the free wheel. Fig. 161 shows this device applied to the hubs, and Fig. 162 shows it applied at the connexion of the axle-halves.

Yet another form of differential is the eccentric and connecting-rod type, Fig. 163, in which each axle end is formed with two eccentrics with throws at right angles, and short connecting-rods attached to these having gudgeon-pins common to each pair and sliding in slots in the differential casing.

Thus in the illustration the short connecting-rods L and are attached to the pin E, the other pair J and K being attached to the pin I. It will be seen that the gear forms a balance-gear, the mechanical efficiency of which is reduced by friction in the eccentrics.

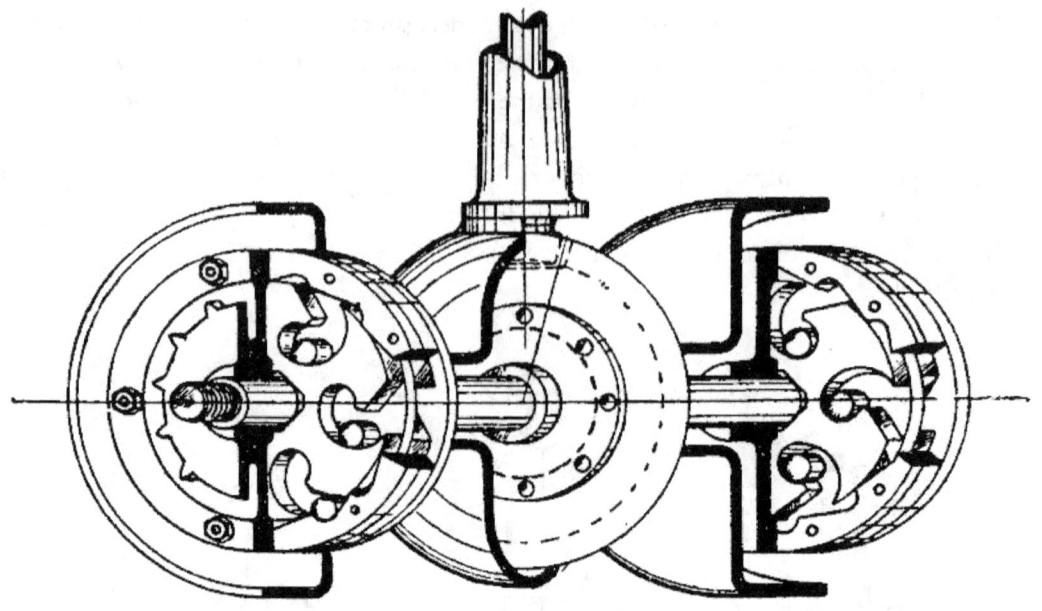

FIG. 161. - *Free-wheel Type Axle.*

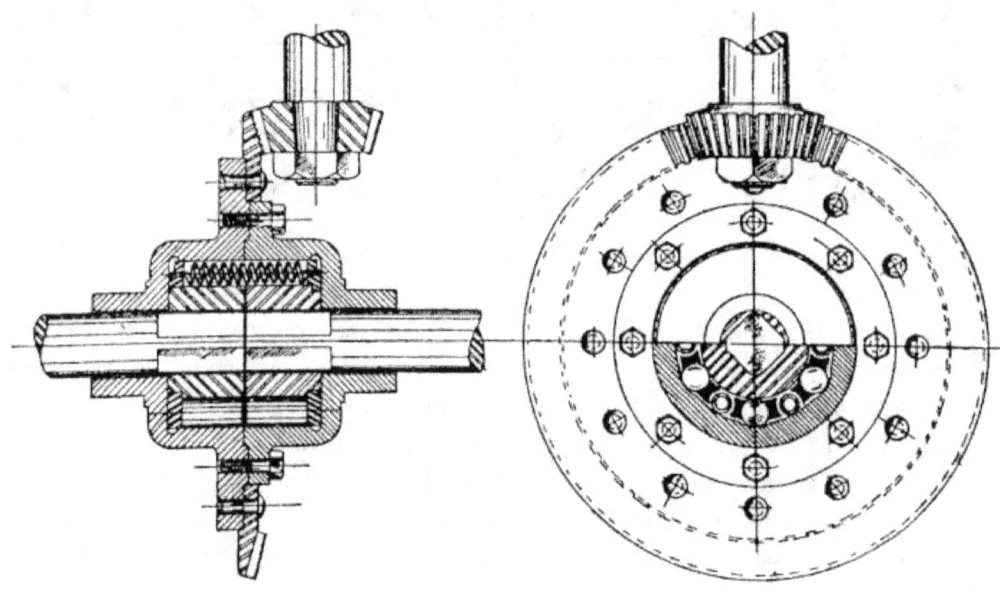

FIG. 162. - *Cam and Roller Type Axle.*

The subject has been dealt with very thoroughly under the title 'Differential Substitutes' in a Paper read by Mr. D. D. Ormsby, before the Society of Automobile Engineers (America) in June 1916.* Mr. Ormsby points out that the trouble found with the ordinary differential is due to its being too efficient, and that it is possible to reduce the reversibility of the gear to any desired extent by using suitably designed worm-gears.

* Transactions of the Society of Automobile Engineers, Vol. xi, 1916, Part II, pages 288-299.

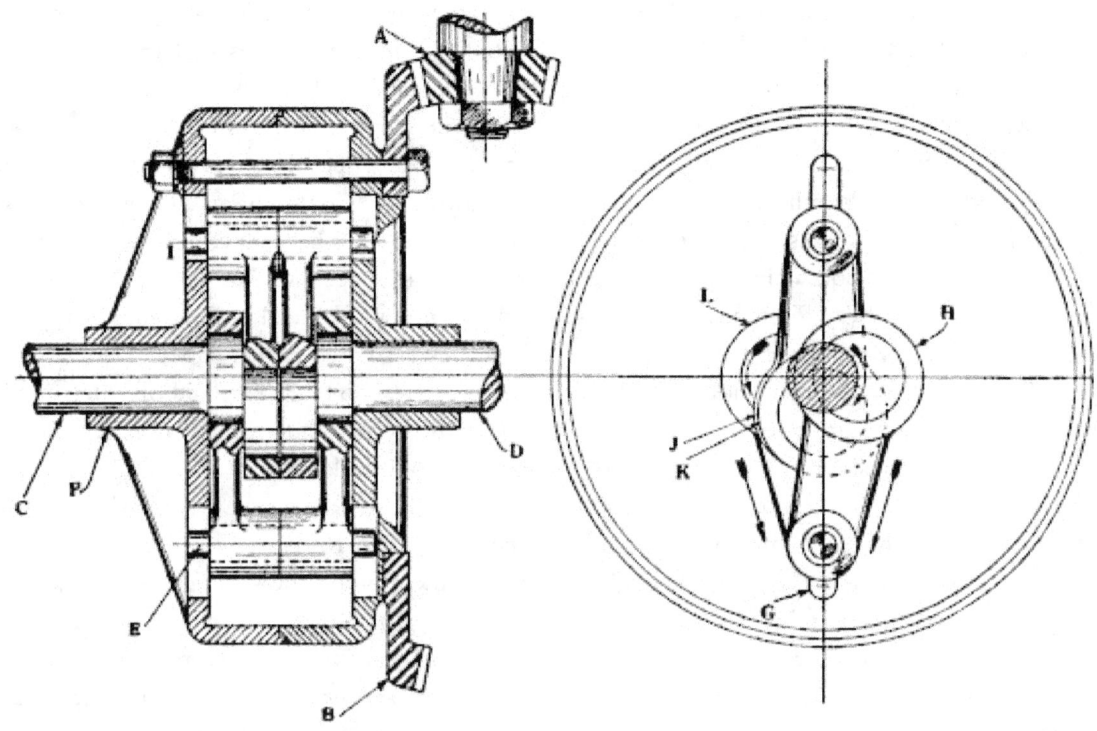

FIG. 163. - *Eccentric and Connecting-Rod Type of Differential.*

The illustrations of worm-differentials, Figs. 5 and 6 (pages 8 - 9), show that the backlash in angular movement of the wheel may be much less than was the case with the Hedgeland axle.

The Author is indebted to this Paper for the illustrations of the Hedgeland free wheel, cam and roller clutch, eccentric and M & S worm differentials.

Like Mr. Ormsby the Author has not come across an example, in practice, of the eccentric and connecting-rod gear shown in Fig. 163.

It was claimed for the Hedgeland gear that a car fitted with it was less liable to side-slip ; the same claim is made for the worm differential gear, but the advantage of the worm differential, which has secured its adoption in one or other of its forms on the majority of four-wheel driven cars intended for use on bad roads or land, is the elimination of difficulty with mud-holes and ice-patches.*

---

* The makers of a model of four-wheel drive car quote the United States Government Specification No. 229 as being fulfilled by their gear, as follows:
"Differential; the differential shall be of such a type as to automatically permit the wheels to revolve at different speeds without complete loss of torque on either wheel, and also to apply at least normal torque to whichever wheel shall have traction. These requirements shall be met when the vehicle is running either forward or backward."

The earliest chain-track invention of which any record has been traced by the Author is that of Richard Lovell Edgeworth of 1770.* In his description the inventor states that this "consists in making portable railways to wheel carriages, so that several pieces of wood are connected to the carriage, which it moves in regular succession in such manner that a sufficient length of railway is constantly at rest for the wheels to roll upon, and that when the wheels have nearly approached the extremity of this part of the railway their motion shall lay down a fresh length of rail in front, the weight of which in its descent shall assist in raising such part of the rail as the wheels have already passed over, and thus the pieces of wood which are taken up in the rear are in succession laid in the front, so as to furnish constantly a railway for the wheels to roll upon." If the word metal is substituted for wood, this description applies to the majority of the types of vehicles described in this Paper.

The next step is the invention of Thos. German in 1801, who placed a chain of rollers carried by linkwork between a runner, or sledge frame, carried on the vehicle, and the ground. This system, modified by the interposition of a chain-track resting on the ground, also has its modern analogues, of which the Log-Hauler and Centiped are the more important. The idea was revived in slightly in 1812, and subsequently by different form by Wm. Palmer Richard Barry in 1821.

In 1825 Sir George Cayley proposed a revolving railway with "the several links or parts . . . connected to each other by joints so formed as to limit the angular movement of any two of the said links relatively with each other by a stop." The form of construction shown in this inventor's drawings closely resembles that adopted on some of the earlier modern attempts to obtain a practical track-laying tractor; and, as has been shown, the chain proposed by him, with its limitation of angular movement, would have been subject to some of the disadvantages that retarded the recent development of this type of tractor in Britain.

The various attempts that were made to produce tractors in which the weight was distributed over a chain of rails or plates failed to evolve any practical device for nearly seventy years, when the less ambitious scheme, invented by Boydell in 1846, of a track-laying wheel, somewhat similar to the girdle wheels now in use for hauling heavy loads over soft land, was carried out in constructional form and actually subjected to severe tests on bad roads about 1858 to 1860. In this tractor the sections or feet were attached to the wheel by links.

Another system, making even less pretence to improvement in load distribution, was the Thomson loose rubber band creeping between the wheel centre and the track of tyre-shoes. In the earliest form the shoes were designed as a continuous chain girdle and later consisted of independent shoes. Road locomotives so fitted were adopted for the traction of road trains on the Grand Trunk road in India under the direction of Colonel R. E. B. Crompton.† The system had an active life of several years in the early seventies. The engines weighed about 9 tons, and hauled some 16 tons of load without undue road wear. The expense of maintaining the rubber tyres under these conditions was as low as 1 5d. per mile-run. The invention constituted the first practical attempt to reduce the unsprung weight of the vehicle to that of the tyre-shoes alone.

---

* Since this Paper was submitted to the Institution a series of articles on the "Evolution of the Chain-Track Tractor" has appeared in The Engineer, Vol. 124, pp. 111-2, 134-6, 156-9, 181-4, 202-5, 221-4, and 241-4.
† Proceedings, I.Mech.E., 1879, page 530.

In the latter half of the nineteenth century but few attempts to revive tracklaying tractors appear to have been made. At the Royal Agricultural Society's show, held at Kilburn in August 1879, two vehicles were exhibited by W. C. Pellatt of Clerkenwell.* These were fitted with self-laying continuous tracks of flat plates secured to chains; the jointing of the successive links was of the most primitive character with practically point contact between the chain links. The resultant wear would be sufficient to account for the failure of these vehicles to attain success although they again demonstrated the advantages of the chain-track for transport over soft ground.

The combination of the two ideas of the chain-track and the foot-spring in the same vehicle was developed by Diplock in his Pedrail alluded to by Dr. Hele-Shaw in a lecture given at Liverpool in 1903,† and in the discussion on Road Locomotion.‡ The diagram given by him, Fig. 40 (page 25), illustrates a further stage in the attempts to improve the springing of track-laying vehicles, a feature considered of but little importance in the earlier forms, but one the value of which is now being appreciated to an increasing extent by the manufacturers of chain-track tractors.

* *The Engineer*, Vol. 47, page 124.
† "The Pedrail," a Revolution in Mechanical Locomotion. Lecture delivered to the Liverpool Self-propelled Traffic Association, by Prof. H. S. Hele-Shaw, at Liverpool University, 28 Nov. 1903.
‡ Proceedings, I.Mech.E. 1910, pages 1561-6.

APPENDIX II.

APPENDIX II.
Authorities Consulted.

Young, C. F. T. The Economy of Steam Power on Common Roads, London, 1860.
Proceedings of the Institution of Automobile Engineers.
Proceedings of the (American) Society of Automobile Engineers.
Proceedings of the Institution of Mechanical Engineers.
Engineering: The Engineer: The Autocar: The Implement and Machinery Review.

TABLE 1 (continued on opposite page).

Four-Wheel Driven Vehicles.

-	Make and Description.	Normal Load.	Weight of Truck.	Total Weight.	Front and Back Wheels.	
		lb.	lb.	lb.	diam. in.	width. in.
1	FWD 1½ - ton Model G	3,000	-	-	36	4
2	FWD 3-ton Model B 1915	6,000	6,000	12,000	36	6
3	FWD 5 – 6 ton.	11,000	10,000	21,000	38	6 dual
4	Jeffrey Quad Model 4016	*4,000*	*5,000*	*9,000*	36	5
5	Walter Super-quad 3-ton Model	*6,000*	*6,000*	*12,000*	40	6
6	Walter Super-quad 5-ton Model	*10,000*	*6,000*	*16,000*	40	7
7	Walter Super-quad Tractor	*6,000*	*6,000*	*12,000*	40	4 dual
8	Couple-Gear 3 ½ ton Petrol-Electric HC	7,000	9,000	16,000	36	3 ½ dual
9	Couple-Gear Electric 5-ton Model AC	10,000	*11,000*	*21,000*	36	4 dual
10	Couple-Gear Electric 5-ton Model A	10,000	*11,000*	*21,000*	36	4 dual
11	Couple-Gear Electric 3 ½ ton Model H	7,000	*9,000*	*16,000*	36	3 ½ dual

Approximate or *estimated* figures are shown in italics.

Engine.

-	Make and Description.	H.P.	No. of Cylrs.	Bore. in.	Stroke. in.	1st Speed. gear ratio.
1	FWD 1½ - ton Model G	28.9	4	4 ½	5	-
2	FWD 3-ton Model B 1915	36.1	4	4 ¾	5 ½	35.6:1
3	FWD 5 – 6 ton.	44.2	4	5 ¼	7	46:1
4	Jeffrey Quad Model 4016	28.9	4	4 ¼	5 ½	42.3:1
5	Walter Super-quad 3-ton Model	30.6	4	4 ⅜	6	67:1
6	Walter Super-quad 5-ton Model	30.6	4	4 ⅜	6	67:1
7	Walter Super-quad Tractor	30.6	4	4 ⅜	6	67:1
8	Couple-Gear 3 ½ ton Petrol-Electric HC	40	4	5	5 ½	25:1
9	Couple-Gear Electric 5-ton Model AC	40	4	5	6	25:1
10	Couple-Gear Electric 5-ton Model A	18 h.p. battery of 44 cells 33 plates each				25:1
11	Couple-Gear Electric 3 ½ ton Model H	9 h.p. battery of 44 cells 17 plates each				25:1

(concluded from opposite page) TABLE 1.

Four-Wheel Driven Vehicles.

-	Track.	Wheel base.	Turning Radius.	Width Overall.	Length Overall.	Platform Length.	Chassis Width Outside.	Clutch.
	in.	in.	feet.	in.	ft. in.	ft. in.	in.	
1	56	124	46	-	16 8	11 4	36	multiple disc*
2	56	124	46	70	16 8	11 4	36	"
3	72	148	34.5	88	21 8	13 0	-	"
4	56	124	24	74	16 10 ½	10 0	38	disk
5	64	132	15	82	18 6	11 6	38	leather cone
6	64	156	17.5	82	20 6	15 0	38	"
7	64	108	12.5	79	16 6	9 0	38	expansion cone
8	66	144	13.5	82	18 6	14 0	-	electric
9	66	144	13.5	89	18 6	14 0	-	"
10	72	106	10	89	14 6	14 6	49	"
11	66	106	10	80 ½	14 6	14 6	43	"

* Hele-Shaw.

Speeds, Brakes, etc.

-	2nd Speed.	3rd Speed.	4th Speed.	Reverse Speed.	Foot brakes. No. Diam. Width.	Emergency brake. No. diam. Width.	Water. Galls. (Brit.)	Petrol. Galls. (Brit.)
	gear ratio.	gear ratio.	gear ratio.	gear ratio.	No. in.	No. in.		
1	-	-*	-	-	8 x 1 ¾	13 ¼ x 2	-	-
2	17.8:1	8.9:1	-	36.1:1	10 x 3 ½	15 ¼ x 2 ¾	6.6	21
3	22.5:1	12.1:1	-	55.3:1	10 x 4	15.9 x 4	-	21
4	24.7:1	14.05:1	8.5:1	45:1	4 x 6 ¾ x 2 ½	1 x 8 x 2 ½	9.2	22
5	27:1	13.6:1	8.4:1	51:1	2 x 12 x 4	1 x 10 x 3 ½	-	25
6	27:1	13.6:1	8.1:1	51:1	2 x 12 x 4	1 x 10 x 3 ½	-	25
7	27:1	13.6:1	8.4:1	51:1	2 x 12 x 4	1 x 10 x 3 ½	-	21
8	-	-	-	25:1	4 x 17 x 2	electric	-	17
9	-	-	-	25:1	4 x 17 x 2	"	-	17
10	-	-	-	25:1	4 x 17 x 2	"	-	-
11	-	-	-	25:1	4 x 17 x 2	"	-	-

* Geared to 16 miles per hour.

TABLE 2 (continued on opposite page).

Chain-Track Tractors.

No	Fig.	Pl.	Description.	Maker.	h.p.	No. of Cylinders.	Bore (in.).	Stroke (in.).
1	59	9	Log-Hauler (steam)	Phoenix	100	4	6.25	8.0 d.a.
2	66	11	Centiped 50 h.p.	Phoenix	50	4	5.5	7.0
3	73	-	Tractor-Truck Type A	Allis-Chalmers	68	4	5.25	7.0
4	74	13	Caterpillar 120 h.p.	Holt	120	6	7.5	8.0
5	75	14	Caterpillar 75 h.p.	Holt	75	4	7.5	8.0
6	76	14	Caterpillar 45 h.p.	Holt	45	4	6.0	7.0
7	78	15	Caterpillar 18 h.p.	Holt	18	4	4.5	6.0
8	92	19	Clayton Tractor 11 h.p.	Clayton &	110	6	5.75	6.5
9	100	-	Clayton Tractor 35 h.p.	Shuttleworth	35	4	4.75	5.5
10	101	19	Tracklayer 90 h.p.	C. L. Best	90	4	8.0	9.0
11	-	-	Tracklayer 75 h.p.	C.L. Best	75	4	7.75	9.0
12	102	20	Tracklayer 30 h.p.	C. L. Best	30	4	5.25	6.25
13	-	-	Tracklayer 16 h.p.	C. L. Best	16	4	4.375	5.25
14	-	-	Creeping Grip 'Giant'	Bullock	75	4	7.5	9.0
15	107	21	Creeping Grip 'Senior'	Bullock	50	4	6.5	8.0
16	-	-	Creeping Grip 'Junior'	Bullock	30	4	5.0	7.0
17	108	21	Creeping Grip 'Baby'	Bullock	16	2	6.0	6.0
18	112	22	Austin Tractor No. 35	F. C. Austin	35	4	5.0	6.5
19	113	23	Austin Tractor No. 15	F.C. Austin	15	4	4.25	5.5
20	114	23	Burford-Cleveland Tractor	H. G. Burford	20	4	3.5	5.125
21	119	26	Strait Tractor	Killen-Strait	50	4	4.75	6.75
22	119	26	Strait Tractor	Killen-Strait	50	4	4.75	6.75
23	122	27	Strait Tractor Mod. 3	Killen-Strait	25	4	4.25	5.75
24	124	28	Yuba Tractor Mod. 18	Yuba Ball-tread	35	4	5.25	5.75
25	124	28	Yuba Tractor Mod. 12	Yuba Ball-tread	25	4	4.5	5.5
26	137	33	Martin Cultivator 25 h.p.	Martin	25	4	3.75	5.0
27	147	35	Lefebvre Tractor 40 h.p.	Lefebvre	40	4	5.125	6.25
28	-	-	Lefebvre Tractor 30 h.p.	Lefebvre	30	4	4.375	5.125
29	152	37	Trench Excavator Mod. 60 (steam)	Parsons	80	2	7.5	10.0 d.a.
30	153	37	Trench Excavator Mod. 48	Parsons	62	4	7.25	9.0
31	-	-	Trench Excavator Mod. 36	Parsons	45	4	6.25	8.0
32	154	38	Trench Excavator Mod. 24	Parsons	30	4	5.0	7.5
33	-	-	Trench Excavator Mod. 18	Parsons	20	4	5.0	6.0
34	-	-	Trench Excavator Mod. 15	Parsons	30	4	5.0	7.5

Approximate and *estimated* figures are shown in italics.

(continued on next page) TABLE 2.

Chain-Track Tractors.

No.	Rated r.p.m.	Main Clutch.	Speeds in m.p.h. at normal Engine revolutions.				Capacity: British Gallons.			
			1st.	2nd.	3rd.	Reverse.	Petrol.	Paraffin	Bulk Oil.	Water
1	-*	Throttle	0	To	5.0	0 to 5.0	-	-	-	-
2	800	-	1.5	3.0	5.5	1.5	-	50	-	-
3	1,000	Cone	1.61	3.5	6.2†	1.41	21	-	-	9
4	550	5 disk	2.13	(3.5)	-	2.13	-	53.5	5.0	67
5	550	5 disk	2.13	3.5	-	2.13	-	53.5	5.0	44
6	625	5 disk	1.5	2.13	3.5	1.5	-	36	-	7
7	750	3 disk	1.5	2.2	3.6	1.5	-	12	-	4.2
8	1,000	7 disk	1.25	3.5	5.0	2.0	2	12	5	12
9	1,000	Cone	2.0	4.0	-	1.75	2	12	-	5
10	450	Expanding	1.5	2.375	-	1.625	6	66	7	27
11	450	Expanding	1.5	2.375	-	1.625	6	66	7	27
12	600	Expanding	1.75	2.5	-	2.0	-	21	1.2	-
13	650	Expanding	1.75 to 2.38	-	-	1.75 to 2.38	-	12.5	-	-
14	550	Expanding	1.06	2.4	3.4	1.77	12	43		50
15	600	Expanding	1.06	2.4	3.4	1.77	12	46	-	120
16	600	Expanding	2.25	4.0	-	2.25 to 4.0	12	12		21
17	550	Expanding	2.5	-	-	2.5	12.5	-	-	-
18	800	-	1.25	2.5	3.75	1.25 to 3.75‡	16.7	33.3		47
19	800	-	1.25	2.5	3.75	1.25 to 3.75‡	7.5	15	-	11.7
20	1,000	2 disk	3.2	-	-	3.2	2	6		10
21	850	Cone	2.0	3.0	-	2.0	25	-	2.5	12.5
22	850	Cone	2.0	3.0	-	2.0	25	-	2.5	12.5
23	900	Cone	2.25	-	-	2.25	10	-	2.5	7
24	700	-	1.96	3.05	-	1.96	4.5	22.5	3.2	-
25	700	-	1.64	2.66	-	1.64	3	15	1.3	-
26	1,000	2 disk	3.0	-	-	3.0	7.5	-	-	-
27	900	Disk	2.15	2.92	-	2.15	22	-	-	8.8
28	1,000	Disk	2.15	2.92	-	2.15	22	-	-	6.6
29	175*	Expanding	0.002 to 0.045		1.5	1.5	-	-	-	170
30	500	Band	0.004 to 0.045		1.375	1.375	43	-	-	170
31	600	Band	0.004 to 0.055		1.5	1.5	38	-	-	136
32	650	Band	0.008 to 0.110		1.5	1.5	27.5	-	-	89
33	700	Band	0.008 to 0.110		1.5	1.5	21.5	-	-	39
34	650	Band	-	-	1.5	1.5	27.5	-	-	89

*Steam. †4th Speed 7.5. ‡3 Speeds on reverse.

TABLE 2 (continued from previous page).

Chain-Track Tractors.

No.	Fig.	Pl.	Description.	Maker.	Supports to each Track.	Track Width. Normal	Wide
						in.	in.
1	59	9	Log-Hauler (steam)	Phoenix	14 rollers	12.0	-
2	66	11	Centiped 50 h.p.	Phoenix	14 rollers	12.0	-
3	73	-	Tractor-Truck Type A	Allis-Chalmers	15 rollers	12.5	-
4	74	13	Caterpillar 120 h.p.	Holt	5 wheels	24.0	30.0
5	75	14	Caterpillar 75 h.p.	Holt	5 wheels	24.0	30.0
6	76	14	Caterpillar 45 h.p.	Holt	5 wheels	13.0	30.0
7	78	15	Caterpillar 18 h.p.	Holt	5 wheels	11.0	-
8	92	18	Clayton Tractor 110 h.p.	Clayton &	7 wheels	22.5	-
9	100	-	Clayton Tractor 35 h.p.	Shuttleworth	4 wheels	14.0	-
10	101	19	Tracklayer 90 h.p.	C. L. Best	7 wheels	24.0	30.0
11	-	-	Tracklayer 75 h.p.	C. L. Best	5 wheels	24.0	30.0
12	102	20	Tracklayer 30 h.p.	C. L. Best	4 wheels	12.0	20.0
13	-	-	Tracklayer 16 h.p.	C. L. Best	3 wheels	10.0	-
14	-	-	Creeping-Grip 75 h.p.	Bullock	4 wheels	20.0	-
15	107	21	Creeping-Grip 50 h.p.	Bullock	4 wheels	20.0	-
16	-	-	Creeping-Grip 30 h.p.	Bullock	3 wheels	12.0	-
17	108	21	Creeping-Grip 16 h.p.	Bullock	3 wheels	12.0	-
18	112	22	Austin Tractor 35 h.p.	F. C. Austin	6 wheels	12.0	36.0
19	113	23	Austin Tractor 15 h.p.	F. C. Austin	2 wheels	12.0	-
20	114	23	Burford-Cleveland 20 h.p.	H. G. Burford	3 wheels	6.0	-
21	119	26	Strait Tractor 50 h.p.	Killen-Strait	3 wheels	18.0	24.0
22	119	26	Strait Tractor 50 h.p.	Killen-Strait	3 wheels	30.0	36.0
23	122	27	Strait Tractor 25 h.p.	Killen-Strait	3 wheels	17.0	-
24	124	28	Yuba Tractor 35 h.p.	Yuba Ball-tread	2 x 21 balls	17.0	-
25	124	28	Yuba Tractor 25 h.p.	Yuba Ball-tread	2 x 16 balls	15.0	-
26	137	33	Martin Cultivator 25 h.p.	Martin	3 wheels	8.0	-
27	147	35	Lefebvre Tractor 40 h.p.	Lefebvre	} Not	10.2	-
28	-	-	Lefebvre Tractor 25 h.p.	Lefebvre	} supporting	8.3	-
29	152	37	Trench Excavator Mod. 60	Parsons	6 wheels	26.0	-
30	153	37	Trench Excavator Mod. 48	Parsons	6 wheels	27.5	-
31	-	-	Trench Excavator Mod. 36	Parsons	6 wheels	26.0	-
32	154	38	Trench Excavator Mod. 24	Parsons	6 wheels	25.5	-
33	-	-	Trench Excavator Mod. 18	Parsons	6 wheels	24.0	-
34	-	-	Trench Excavator Mod. 15	Parsons	9 wheels	30.0	-

Approximate and *estimated* figures are shown in italics.

(continued on next page) TABLE 2.

Chain-Track Tractors.

No.	Track Length	Area of Tracks.		Weight (lb.) carried on			Total Weights (lb.).		Insistent Weight in lb. per square inch.	
		Normal	Wide.	Tracks.		Steering Wheels.	Normal	Wide.	Normal	Wide.
	in.	sq. in.	sq. in.	Normal	Wide.					
1	60	1,440	-	30,500	-	5,500	36,000	-	21.18	-
2	60	1,440	-	13,600	-	3,200	16,800	-	9.45	-
3	60	1,500	-	14,400	-	3,600	18,000	-	9.60	-
4	80	3,840	4,800	23,750	-	2,500	26,250	-	6.19	4.95
5	80	3,840	4,800	21,350	-	2,250	23,600	-	5.56	4.45
6	70	2,080	4,800	13,000	14,500	-	13,000	14,500	6.25	3.02
7	64	1,408	-	6,000	-	-	6,000	-	4.26	-
8	100.5	4,522	-	26,880	-	3,360	30,240	-	5.94	-
9	71.3	2,000	-	5,040	-	-	5,040	-	2.52	-
10	90	4,320	5,400	25,000	25,000	5,000	30,000	30,000	5.79	4.63
11	66	3,168	3,960	23,300	23,300	4,700	28,000	28,000	7.36	5.88
12	66	1,584	2,640	8,500	8,500	-	8,500	8,500	5.37	3.22
13	66	1,320	-	5,100	-	-	5,100	-	3.86	-
14	64	2,560	-	19,500	-	3,500	23,000	-	7.62	-
15	64	2,560	-	16,500	-	3,000	19,500	-	6.45	-
16	48	1,152	-	6,500	-	1,500	8,000	-	5.64	-
17	48	1,152	-	5,200	-	-	5,200	-	4.51	-
18	96	2,300	6,900	10,800	12,300	-	10,800	12,300	4.70	1.78
19	60	1,440	-	6,750	-	-	6,750	-	4.70	-
20	50	600	-	3,000	-	-	3,000	-	5.00	-
21	48	1,728	2,304	8,000	8,000	1,500	9,500	9,500	4.63	3.47
22	48	2,880	3,456	8,500	8,500	1,500	10,000	10,000	2.95	2.46
23	48	816	-	3,000	-	1,000	6,000*	-	3.68	-
24	48	1,632	-	7,400	-	1,500	8,900	-	4.53	-
25	36	1,080	-	6,000	-	1,600	7,600	-	5.55	-
26	40	640	-	3,250	-	670†	3,920	-	5.01	-
27	63	-	-	-	-	4,000	7,050	-	-	-
28	51	-	-	-	-	1,780	5,950	-	-	-
29	72	3,744	-	27,000	-	27,000‡	54,000	-	7.21	-
30	72	3,960	-	33,800	-	11,200	45,000	-	8.54	-
31	66	3,432	-	22,000	-	8,000	30,000	-	6.41	-
32	62	3,162	-	12,000	-	5,000	17,000	-	3.80	-
33	41	1,962	-	7,000	-	2,600	9,600	-	3.56	-
34	78	4,680	-	20,000	-	-	20,000	-	4.27	-

* 2000 lb. on wheel. † On ploughs. ‡ Two pairs of wheels.

TABLE 2 (*continued from previous page*).

No.	Fig.	Pl.	Description.	Maker.	Supports to each Track.
1	59	9	Log-Hauler (steam)	Phoenix	Runners and engines
2	66	11	Centiped 50 h.p.	Phoenix	Front wheels
3	73	-	Tractor-Truck Type A	Allis-Chalmers	Front wheels
4	74	13	Caterpillar 120 h.p.	Holt	Front wheel and clutches
5	75	14	Caterpillar 75 h.p.	Holt	Front wheel and clutches
6	76	14	Caterpillar 45 h.p.	Holt	Clutches and brakes
7	78	15	Caterpillar 18 h.p.	Holt	Clutches
8	92	18	Clayton Tractor 110 h.p.	Clayton &	Front wheel and brakes
9	100	-	Clayton Tractor 35 h.p.	Shuttleworth	Clutches and brakes
10	101	19	Tracklayer 90 h.p.	C. L. Best	Front wheel and clutches
11	-	-	Tracklayer 75 h.p.	C. L. Best	Front wheel
12	102	20	Tracklayer 30 h.p.	C. L. Best	Clutches and brakes
13	-	-	Tracklayer 16 h.p.	C. L. Best	Clutches and brakes
14	-	-	Creeping-Grip 75 h.p.	Bullock	Front wheels
15	107	21	Creeping-Grip 50 h.p.	Bullock	Front wheels
16	-	-	Creeping-Grip 30 h.p.	Bullock	Front wheel
17	108	21	Creeping-Grip 16 h.p.	Bullock	Clutches*
18	112	22	Austin Tractor 35 h.p.	F. C. Austin	Clutches and brakes
19	113	23	Austin Tractor 15 h.p.	F. C. Austin	Clutches and brakes
20	114	23	Burford-Cleveland 20 h.p.	H. G. Burford	Band brakes
21	119	26	Strait Tractor 50 h.p.	Killen-Strait	Front track
22	119	26	Strait Tractor 50 h.p.	Killen-Strait	Front track
23	122	27	Strait Tractor 25 h.p.	Killen-Strait	Front track
24	124	28	Yuba Tractor 35 h.p.	Yuba Ball-tread	Front wheel*
25	124	28	Yuba Tractor 25 h.p.	Yuba Ball-tread	Front wheel*
26	137	33	Martin Cultivator 25 h.p.	Martin	Dog clutches
27	147	35	Lefebvre Tractor 40 h.p.	Lefebvre	Front wheels
28	-	-	Lefebvre Tractor 25 h.p.	Lefebvre	Front wheel
29	152	37	Trench Excavator Mod. 60	Parsons	Front and back wheels
30	153	37	Trench Excavator Mod. 48	Parsons	Front wheels
31	-	-	Trench Excavator Mod. 36	Parsons	Front wheels
32	154	38	Trench Excavator Mod. 24	Parsons	Front wheels
33	-	-	Trench Excavator Mod. 18	Parsons	Front wheels
34	-	-	Trench Excavator Mod. 15	Parsons	Clutches

Approximate and *estimated* figures in italics.   * Either track be reversed.

(*concluded from opposite page*) TABLE 2.

Chain-Track Tractors.

No.	Tractor overall.		Turning radius	Drawbar pull on low gear.	Front Wheels.			Pitch of chain-track.	Drawbar pull. Per cent of weight.	
	Normal width	Length			No.	Width	Dia.		Normal	Wide
	ft. in.	ft. in.	ft. in.	lb.		in.	in.	in.		
1	5 4	27 6	75 0	-	2	runners		9.5	-	-
2	5 4	21 0	25 0	8,000	2	10	40	9.5	47.6	-
3	5 6	21 6	26 0	8,000	2	10	40	7.6	44.4	-
4	8 8	21 0	16 0	12,000	1	18	36	9.25	45.7	-
5	8 8	19 6	15 0	8,500	1	18	36	9.25	36.0	-
6	6 8	12 9	6 0	4,500	None	-	-	9.25	34.6	31.0
7	4 3	9 6	6 0	2,800	None	-	-	7.5	46.7	-
8	7 4	17 6	10 0	22,000	1	6	50	7.75	72.8	-
9	5 4	10 10	7 0	-	None	-	-	7.5	-	-
10	8 9	24 6	-	14,500	1	20	40	9.75	48.3	-
11	8 7	22 4	-	12,000	1	24	48	9.75	42.9	-
12	5 5	9 4	6 0	3,500	None	-	-	7.25	41.2	-
13	4 2	7 0	5 0	1,750	None	-	-	4.6	34.3	-
14	8 0	18 8	30 0	13,000	2	10	40	9.0	56.5	-
15	8 0	18 0	30 0	8,200	2	10	40	9.0	42.1	-
16	5 8	13 0	22 0	3,330	1	14	32	7.0	41.6	-
17	5 3	11 6	8 0	-	None	-	-	6.0	30.6	-
18	5 8	10 0	6 0	5,400	None	-	-	9.0	50.0	43.9
19	6 0	9 0	6 0	3,000	None	-	-	9.0	44.4	-
20	4 2	8 0	6 0	2,000	None	-	-	8.0	66.7	-
21	6 0	13 0	8 0	5,600	1	14	} Chain track	6.5	59.0	59.0
22	8 0	13 0	8 0	5,600	1	14		6.5	56.0	56.0
23	5 10	12 10	8 0	2,500	1	12		6.5	41.7	-
24	6 1	15 1	7 6	4,600	1	14	36	3.75	51.7	-
25	5 9	14 7	7 9	3,400	1	14	36	3.75	44.7	-
26	4 6	16 0†	8 0	1,900	None	-	-	7.0	48.5	-
27	6 1	18 0	16 6	4,400	2	10	30	8.5	62.4	-
28	5 9	15 8	1 6	3,100	1	14	30	8.5	52.0	-
29	10 0	35 0	30 0	-	2‡	14	40	7.0	-	-
30	10 5	46 6	50 0	-	2	20	40	6.0	-	-
31	8 9	36 6	50 0	-	2	20	34	6.0	-	-
32	7 8	28 3	30 0	-	2	15	30	6.0	-	-
33	7 0	23 0	30 0	-	2	15	26	5.0	-	-
34	10 0	15 0	30 0	-	none	-	-	6.0	-	-

†Includes ploughs. ‡Also two back wheels.

Discussion.

Mr. H. G. BURFORD wished to congratulate Mr. Legros on the very excellent Paper or Lecture which he had delivered at the last Meeting. The general question relating to road traction and the points of historical interest in the Paper had been dealt with many times, both before this Institution and before the Institution of Automobile Engineers. Referring first to Part I of the Paper, one of the interesting features was the detail given of the four-wheel drive. This drive, as far as its use for commercial purposes in this country was concerned, had been developed rapidly since the country had been engaged in war. In his opinion, however, the commercial usefulness of the four-wheel drive was very limited in settled countries such as Great Britain, but there was certainly a very large field for it in new countries where roads were not as good as they were here.

The Author had not pointed out, as he could well have done from his large experience, the principal features that engineers should have in mind when designing and constructing a commercial vehicle to meet modern requirements. From his own experience he much preferred a machine or commercial vehicle built on the unit principle - for example, the engine and transmission as one unit, the rear axle and differential as another unit, frame and springs another, front axle and steering another - so that each unit was self-contained. This greatly reduced the cost of operation. Instead of the usual repairing of any part on the chassis, where large quantities of vehicles were employed, new units could be put in immediately for replacemen.t. The great question in connexion with commercial use was that of interchangeability. The Author had given a historical review of the various types, but he might have laid more emphasis on the question of reliability, interchangeability, and the principles of design which would enable quick and cheap replacements.

Mr. Legros had also drawn attention to the question of various methods of locking the differential gear. From his own past 18 years' experience of ordinary commercial cars, nearly all the designs he had studied and worked had been of very little practical use, the reason being that if an ordinary drive got into a position in which it could not be got out with one wheel, it was equally hopeless when the other wheel was locked. He respected the Author's authority on motor traction in every way, having known him for nearly 22 years in connexion with the movement, and he endorsed very fully his remarks on that subject – that some intelligent investigation was required on some points and particularly the locking of differentials, and this might well receive the support of the Institution.

Another question relating to commercial vehicles, which had been discussed both in this country and in America, was the final drive or final form of transmission. Apparently this had developed under two heads:- (1) the worm drive, and (2) the internal pinion or rack and pinion. In studying the development of the industry in America, it was really remarkable to note the progress that internal gear-driven machines had made in that country. From his own experience he had always been, and still was, a very strong advocate of the internal rack and pinion drive as against the worm drive which was so common in this country.

With regard to braking, the Author gave some information as to the application of brakes to commercial vehicles, and suggested that four wheels could be used for braking instead of the usual practice of braking on, the two rear-wheels. Admitting that the efficiency would be higher, he thought the steering, the cost of manufacture, and the many difficulties which surrounded the application of front-wheel brakes, would not justify the expense incurred. Further, having regard to the general improvement in motor driving in this country, and the fact that the average person had a better idea of speed than formerly - and, after all, it was a question of speed and judgment - he thought the ordinary commercial vehicle was doing quite well with the back-wheel brakes as now used.

In his opinion Part II was the most interesting part of the Paper. It set out in detail the various forms of chain-track in use, whether actually of chain-driven types, pedrails or other designs, apart from the ordinary wheel-driven machines. It was probable that the war, more than anything else, had brought into prominence the application of the chain-track to the 'Tanks' more than would have been the case in two or three hundred years of peace. Various types of machines had been put forward by the Author for consideration, but he would like to have had some guidance from an expert, a gentleman who had studied the question very carefully and closely, as to the best means of control, the best means of steering, and the approximate weight he would consider with the appropriate width of track to satisfy the agricultural interests of this country. At the moment the agricultural question was uppermost in most men's minds, seeing that we now had to rely so much in this country on agriculture for our daily bread. The value of the Paper and of Papers of a similar description would be considerably increased if they could assimilate the experience of engineers who had already manufactured various types of vehicles. If it were possible for the Author to read a Paper dealing with the subject of why various types of vehicles or tractors had become obsolete, and why they were discarded, whether on the grounds of cost of manufacture or on the ground of inaccessibility or for any other reason, it would be of great value. Of course it would cost considerable research and money. The reason why he suggested that was because it would save so much of the time of young engineers and firms who were entering the. engineering business of manufacturing, if they could ascertain why such and such a thing had failed - whether it failed from constructional defects, faulty material or from the wear and tear of actual use. If a Paper of that kind were published in the Proceedings of the Institution, the world generally, and engineers in particular, would have a good starting point that would be helpful in the future.

Colonel R. E. B. CROMPTON, C.B., R.E. (T.), said he rose with great pleasure to say a few words on the Paper, for which he felt partly responsible, as he believed the idea of it originated when Mr. Legros was working with him on a very important national matter connected with the subject, of which they were not now allowed to speak. To his mind the value of the Paper lay in drawing attention to the importance of applying mechanical power to haulage purposes in situations off the highways, situations such as were familiar on the war front in a greatly accentuated degree; for it could hardly be called ground that the vehicles had to travel over; it resembled more a morass in a partly liquid form. These difficult requirements had undoubtedly stimulated invention, and improved design would be a means of meeting the smaller difficulties of the same class which the farmer encountered when he utilized mechanical haulage for ploughing his land or for removing his produce from fields without over - compressing or damaging its surface. He congratulated the Institution on having received from the Author such a valuable compilation of what had already been done, as it could not fail to be of great use to designers who were now engaged on this work. Whether it could be further supplemented, as Mr. Burford had suggested, he very much doubted, for however desirable, it would be a long and difficult task; but whether supplemented or not, the Paper was an extremely valuable one.

As an old man, who had bridged a good span of life, he wished to try to do justice to those who had worked on this problem in the past.

It was probably forgotten by many of the members that he himself had read a Paper before the Institution on Thomson's road engines,* used by him in India for traversing soft ground; this was thirty-nine years ago, and he had still a good many copies left, because during the twenty years which followed that Paper the whole matter, which was then thoroughly (developed and in a fair way to do what was being done now, went to sleep on account of the determined opposition, not of the road authorities but of railway companies, who thought that the railway and its development, either as light railways or tramways, was the only way of extending the locomotion of a country. Everyone knew now what a mistake that had been, and it could be seen how completely any system of railways, not supplemented by a system of road carriage, failed to develop the farming interest of a country, failed to distribute its food, and had now brought it to the brink of famine.

* Proceedings, I.Mech.E. 1879, page 494.

When in the Army as a mechanical engineer with a turn for studying road locomotion, having constructed vehicles of his own with his own hands, he was sent home by the Government of India, and entrusted with a considerable grant of money to purchase and work the late R. W. Thomson's great invention of road steamers. Road steamers were connoted with the invention of rubber tyres, first with the then little known pneumatic pattern, which, on account of difficulties of manufacture, did not at first develop, but by his successful development of driving wheels with large flat surfaces on the ground, thereby extending their surface and enormously increasing the tractive power of the tractor. He was sent home from India with the object of developing these engines, so that they could be used in India for carriage on the existing roads and across open country, and so do away with the necessity for the extension of light railways for collecting crops. These vehicles could be used in one part of the country for one season of the year, and might be transferred to another part of the country at another time of the year, which was peculiar to all systems of road transport. He found Mr. Thomson the mechanical genius of his time - 1870. His chief draughtsman and designer was the present Sir Alexander Kennedy, and he believed that Sir Alexander Kennedy, Mr. Richard Muirhead, and himself were probably the only people alive now who saw much of or had the privilege of enjoying conversations with R. W. Thomson. R. W. Thomson foresaw everything, indeed much that was foreshadowed in Mr. Legros's Paper - the Boydell girdle, which has been adopted by the Germans and by ourselves for guns in this war; the further development of various forms of chains, in Thomson's case, encircling the rubber tyre were all discussed and many of the difficult points that Mr. Legros had so ably brought forward were foreseen by Thomson. He was sure that Sir Alexander Kennedy would agree with him in that.

All those who had worked on the 'caterpillar' problem, in order to get a very extended flat foot in contact with the ground, knew that the principal wear took place at two points, at the pin-joints where there was the greatest angular motion, that is, at the point where the chain came down from the leading wheel on to the ground surface. There was little motion on the pin during the time it was passing over the flat surface of the ground, but when it was lifted up at the back of the track there was another large angular motion on the pin, and all experience had been to go back as nearly as possible to the original wheel, that was, to make both the front wheel and back wheels of the tracks of very large diameter.

He was not able to say very much on this subject, because it was closely connected with developments in the Field. The limitation of angular pin movement was one of the things to which attention had to be directed. It was necessary to get the happy mean between a short track with a short flat foot, which was almost the same as a very large wheel, and the other extreme, the very long track with a very large angular movement at each end.

Then came the second difficulty, that of taking up the slack due to the chain wear at the pin-joints. Thomson produced his flat foot by a very thick rubber tyre which practically gave an endless chain; of course, this was really supplying an infinity of pin-joints within the rubber itself, and which he thought would not wear or give any trouble, but unfortunately internal wear took place in the rubber; what electrical engineers called hysteresis took place - that resistance to internal molecular motion which was developed as heat at points of acute flexure and destroyed the rubber. The same thing occurred every day in motor-'bus tyres; owners of those vehicles had recently called upon him to take up the matter so as to tell them all that happened when they use a material like rubber to take the place of a mechanical chain. A Committee had been appointed to deal with it after the war.

In Thomson's case the flat foot was obtained by an enormous expenditure on rubber, his tyres being 4 ½ inches thick, and the rubber was found to wear on both surfaces. On the interior surface the wheel slipped inside the rubber and abraded that surface. Mr. Legros had mentioned the reason, which was that some slipping had to take place at that point instead of the differential, thus allowing a friction-clutch inside the rubber tyre.

This was an expensive mistake, and had to be abandoned. The outside of the rubber tyre was discovered to have a very low coefficient of adhesion in certain conditions, such as greasy surfaces of land or road, and it was necessary to supplement it with a steel chain on the outside. Then at once came the 'caterpillar' difficulties, the perpetual stretching and lengthening of the chains due to the wear of the pin-joints. They became so slack that they had to be shortened on the road, and it was necessary to carry an arrangement of cramps and mechanism for clipping them up and cutting out a length and joining them together again. As he was one of the men who were developing the thing, he was one of those who came in for much of the abuse. To-day it was possible to do a great deal better in the utilization of chains, owing to the wonderful development of roller chains by various makers, such as those by Hans Renold, the Coventry Chain Company, and others. The magnificent workmanship rendered possible by automatic tools allowed roller chains to be turned out of the very highest class of material, case-hardened all over, and with very small play to allow mud and dirt to get in. This was very hopeful. Now that the 'caterpillar' makers had had the advantage of the knowledge of these things, it was hoped that by sheer good workmanship a great deal of the troubles experienced in the use of those kinds of chains in those early days would be avoided. He had touched on a good many of the points, and what he wished to impress upon the members was that there never was a time when a mechanical engineer who had in him the art of design could address himself to a more worthy object than the production of a thoroughly efficient, well-conceived tractor that the British farmer could use on his land or for bringing his produce off the fields to market. In justice to Mr. Burford, he ought to say that he was glad to find that one of the finest designs that he had seen of that kind was by Mr. Rollin White, the inventor and developer of the White steam-car, who was well known as a real mechanical genius. His ploughing tractor, which he had examined, was the most beautiful example of simplicity and brilliant ideas brought together and worthy of study. If the other forms of ploughing tractors mentioned in the Paper were anything like Mr. White's, he congratulated the world on the progress that it had made in mechanical design as applied to so worthy an object.

MR. GEORGE WATSON said, with regard to Mr. Legros' Paper, he did not think his friend Mr. Burford quite allowed for the difficulties Mr. Legros was under in giving such an extraordinarily comprehensive survey of the work of very many engineers all working more or less in rivalry. The Paper was to a great extent a model of what a Paper ought to be in describing rival schemes and arrangements, without hurting anybody's feelings. For his part he regarded the Paper as a standard work of reference, and he thought the Author had given a most valuable record of his experience without committing himself too far to opinions on a subject which was still very much open to discussion. He had said little or nothing about steam-tractors as applied to the various purposes, but there was no doubt the steam-tractor would still be heard of a good deal. Mr. Clarkson's coke-fired steamers were very likely to find many applications before long.

Mr. Legros mentioned one of the difficulties that perhaps was the chief difficulty of tractors on land, when he spoke of the nut-cracker action which took place in the chain-track mechanism when clay got into the interstices, and of course sooner or later brought stones with it. Unfortunately under those circumstances it was generally the nut-cracker that was cracked. He had seen a good many tractors lately and he thought that most of them were not sufficiently guarded against picking up stones and dirt. Even where stones did not get in, the wear and tear on the gear was enormous, and the pinions wore out very rapidly, eventually breaking from weakness if not replaced.

The question that interested him most at the present time, as it did so many people, was the agricultural tractor to which Mr. Amos's Paper was an important contribution. He had no doubt at all that the use of the agricultural tractor was going to be one of the chief features in defeating the U-boat, and he believed it would be an even more important factor if administration kept pace with the efficiency of engineering.

The question of speed in cultivation had been dealt with by Mr. Amos in connexion with the farmers' problems, and he thought that was of enormous importance in this climate. If a farmer had a machine which could do his ploughing very rapidly, he was able to take advantage of the few fine days obtained in this country and get the work done at the proper time of the year, when the weather was right. There was no doubt that ploughing in bad and rainy weather was worse than useless, and the motor-tractor gave every farmer an opportunity of doing the work in favourable times.

He did not quite follow Mr. Amos when he said there were certain states of heavy clay land when it was too hard for the motor-tractor to tackle. He had never experienced that, and he thought that, given a reasonable chance, a motor-tractor would tackle any hard land. He had seen a tractor ploughing 7 inches deep on clay which was baked very hard and which had been undisturbed for two years and was full of clover roots; it was ploughing it at the rate of over half an acre an hour, at an expenditure of about 3 ½ gallons of paraffin per acre. It was working with two furrows, and he had no doubt at all that if that ground had been still harder, had it been possible, the work could have been done, though more slowly, with one furrow.

Mr. Amos had said that the question of speed of the plough did not affect very much the expenditure of power required to plough a given area. He did not quite agree with that. There was a certain speed of plough which was the most advantageous. Mr. Amos spoke of the horse-drawn plough going about 2 miles an hour and the motor plough 3 ½ miles or faster, but he believed eventually it would be found that motor-ploughs would be driven at about 2 ½ miles per hour, and that rate would be found to be quite fast enough. The problem was not purely one of friction but of the separation of the earth, and if it was attempted to be done too quickly it was bound to take more power, and therefore to use the power of the engine less efficiently. There was another point about very rapid ploughing it broke up the furrow and left the work in an untidy state with green stuff showing. When ploughing more slowly, a longer breast of plough could be used and the furrow turned clean over, and work could be done resembling that which was done in a ploughing match. The question of weight on the land was of enormous importance. The mistake had been made in many cases of sending a motor which was thoroughly suitable for hard land on to soft boggy land or sandy land. Such motors dug themselves in and were not able to tackle the work. In such circumstances there was nothing better than a chain-track tractor to deal with the question.

Mr. Amos had not dealt with one point which seemed of immense importance in the design of agricultural motors, and that was the question of the position on the land on which the motor worked. If it were working entirely on unploughed land, both the wheels being on the level, the tractor could be provided with ordinary axles without any means of adjusting the level, but the drawback to this was that the pull of the draw-bar was always diagonal, and if the land were soft it tended to drag the tractor into the furrow continually, and also tended to drag the plough out of the straight line. He had seen much better work done with tractors that had two wheels in the furrow. They guided themselves so well that he had seen the man in charge get off the tractor and walk beside it, smoking his cigarette, while the tractor went on and did its work. The drawback to having two wheels in a furrow was that it was only possible to have the driving wheels of a certain width, but, given ordinary hard land, he did not think that that presented any real difficulty.

One other point Mr. Amos did not mention which he believed would come to the front with motor traction, and that was the question of subsoiling. A great deal of land would be immensely benefited by having the subsoil disturbed and broken up below the surface of the plough, but it was most important that it should not be brought to the top. Deep ploughing did not, meet the case at all. He believed that subsoiling with a cultivator tyne of some sort underneath the ploughshare would become a matter of great importance, and, given sufficient power in the motor to draw the machine, the results would be of very great benefit.

MR. WILLIAM H. PATCHELL (Member of Council) said that on the previous Friday Mr. Legros had read his Paper in Lincoln, where it was highly appreciated by a Meeting of about 200. There was no time for discussion, but after the Meeting there was an informal discussion of which no report would appear, so he would like to mention one point that was brought forward by a speaker who had been in East Africa. Mr. Legros showed two slides, Figs. 15 and 16, Plate 3, of the F.W.D. motor-truck, one carrying ore over sand and the other crossing a wide expanse of alkali-mud, where there was a very heavy road. The gentleman from East Africa said he had been very severely taxed by the difference in track of some of the American motors, and thought it was highly important that there should be some sort of standardization of the track, as Mr. Burford had mentioned.

Following Mr. Watson, one was glad to feel a breath of the country in listening to an old mechanical-implement maker speaking; it took one back again to the olden days. Mr. Watson had mentioned the ploughing by tractors of half an acre an hour. In Hereford the week before last, the Ross tractor ploughing unit held the record with an average per tractor per week of just upon 30 acres. When it was remembered that an acre was what one man could plough in a day, tractor-ploughing showed a great advance. In south Lincoln and in north Cambridge he had known cases where the tractor put across the land was not suitable; its tracks could be seen quite plainly in the following year's crops, and that was not acceptable to a farmer.

Colonel Crompton had mentioned, the old Thomson machine. At the time he (Mr. Patchell) was an apprentice he helped to make some and to repair others, and it was not long ago that he had turned up a piece of the old rubber tyres out of a scrap-box. If he remembered rightly, those rubber tyres were 10 or 11 inches wide and 2 ½ inches thick.

Colonel CROMPTON said they went up to 15 inches wide and 5 inches thick.

Mr. PATCHELL said the fascination of that road steamer was so great that an enterprising showman called one day at the works to see if he could have one made in the form of a dragon, capable of wagging its tail, with the exhaust coming out of its mouth! For the reason that Colonel Crompton had mentioned, that idea had to be abandoned!

Major E. G. BEAUMONT, A.S.C. (M.T.), said there was only one feature in Mr. Legros' Paper he wished to refer to, and that was connected with the irreversible type of differential gear illustrated on pages 8 and 9 of the Paper, and to which particular reference was made on page 16. Mr. Legros stated there that, by the use of that form of differential gear, the third differential on the main shaft could be dispensed with. He rather gathered from the statement in that paragraph that it was not because the machines referred to were four-wheel driven machines and four-wheel steering machines, but because the worm type of differential gear did in fact enable a third differential to be dispensed with. It was easy to understand, with the worm type of. differential gear that, although one wheel of the pair on an axle might be actually off the road or slipping, the full torque might be delivered to the wheel which was on the road; but he could foresee that the two wheels on one axle might together have a higher velocity than the pair of the wheels on the other axle, and that therefore between those two axles there still required to be a differential gear if differences of average velocities were to be provided for, as, for example, when a four-wheel driven, two-wheel steering, machine traversed a curve or when there was difference of diameter of the wheels, as sometimes occurred when one set of tyres was partially worn and the other set of tyres was new and of full diameter. The cases quoted of the Jeffery and the Walter quads which employed two differentials were interesting, because for a given radius of turning on a curved path the angularity of the differential joints giving motion to the wheel-driving shafts was very much less than it would be otherwise. That was to say, the use of the four-wheel steering arrangement did reduce the severity of the work on the universal joints, and moreover also assisted in dispensing with the third differential gear by limiting the difference of average velocity of the two axles.

His experience in the working of those four-wheel vehicles was limited, but, he had seen the Jeffery and the F. W. D. in service, and he was aware that with those which had only two differentials, wear and tear of universal joints and other parts which felt the stresses in transmission were rapid as compared with others which used the three sets of differential gears. He would be obliged to Mr. Legros if he would explain further the statement with regard to the sufficiency of two as compared with three differentials.

Professor P. M. BAKER said there were three questions he would like to bring forward on Mr. Legros' Paper. Referring to the electric drive - mentioned only incidentally by the Author, he said that it was possible to arrange for the electric drive to overcome the difficulty met by irreversible differentials, and it seemed to him that the electric drive would offer great advantages. The fact that it could be made practically to take the place of all the differentials and all the universal joints, and do away with the very awkward driving conditions, would seem to give it a very strong claim in the case of the four-wheel driven vehicle. It was quite possible to arrange for an electric drive - either by means of motors with a considerable shunt field, or by a suitable arrangement of series fields, so that all the effects of the irreversible differential could be obtained. It would even be possible to secure that the maximum torque was exerted on the side which was getting the greatest grip, and this without any attention from the driver.

On looking at some of the vehicles of the chain-track type in Lincoln, it seemed to him that the steering wheel out in front, with a sector and screw to adjust the pressure on the ground surface, was adding another little worry to the steersman, and that probably by some mechanical arrangement, which would avoid the necessity of having to worry the driver with a further appliance to look after in steering, it would be possible to obtain a distinct advantage.

Perhaps Mr. Legros would be good enough to explain how far the method of steering by means of the chain-track was successful, and whether, therefore, the steering-wheel arrangement would not ultimately disappear.

One point which his agricultural friends had been rather insistent upon was, that, they said it, was all very well for an engineer to speak about the pressure in lb. per sq. inch on the track, in the case of a chain-track tractor, being only so much - less, in fact, than the pressure on the sole of one's foot; but what about the total load distributed over the area - did it not flatten down ridges? It would be rather interesting to know what effects were thus produced in ploughing. Did the tractor ever get across the furrows, and, in doing so, did the ridge get flattened down in an undesirable way by the chain-track tractor? That was hardly a point that Mr. Legros had dealt with, but he thought Mr. Amos might be able to give some information on the subject.

MR. L. A. LEGROS, in reply, thanked the Members very much for the extremely kind way in which they had received the Paper. With regard to the remarks of Mr. Burford on the principles underlying the design of commercial vehicles, he was entirely in agreement with him when he spoke of the advisability of adopting the unit principle, but it was necessary that any unit principle of construction should be accompanied by easily operated and easily accessible means of connecting the different units together. The engine must, be easily removable from the frame without dismantling much of the transmission gear; the gear-box must be able to be taken out by itself without stripping the chassis of the axles or of the engine, and the same must apply to both the front axle and the rear axle. The axles considered as units should be accompanied by certain parts of the brake-gear, but each axle should be removable independently as a unit.

Figs. 10 and 14, Plates 1 and 2, illustrated units of the type that could be taken down, and the brakes were of the type Mr. Burford had mentioned.

The subject of the final transmission was alluded to in a number of Papers read before the Institution of Automobile Engineers* relating to questions of gear-box and axle.

One of the American four-wheel drive cars was fitted with the internal rack, as shown in Fig. 22 (page 16). Front-wheel brakes were not in his opinion very necessary or valuable except so far as they made the front and back wheels similar in arrangement. It was very desirable that they should be capable of being operated independently.

With regard to the means of steering chain-track tractors, those tractors of lower power which were without front wheels were quite easily controlled by clutches and brakes, and it was usual to lift the front wheel of the Clayton tractor for making sharp turns. The value of the Paper would certainly have been increased if the reasons which caused the obsolescence of vehicles could have been given. Figs. 40 and 52 (pages 25 and 34) and examples of things that had come and gone, but it would obviously be very unwise to criticize any design that might be going through the period of failure at the present time.

The crossing of land for agricultural work had been alluded to by Colonel Crompton, who had also referred to the road steamers of 1879. The wear which took place between the rubber and the track-chain was, of course, very great in those early cases, more particularly because the conditions suitable to the flow of rubber were then not well understood.

* Proceedings, Inst. of Automobile Engineers: Transmission of Power, by L. A. Legros, Vol. iii, page 335; 'Live Axles for Commercial Vehicles,' by Geo. W. Watson, Vol. x, page 161; 'From Engine to Axle,' by Major B. W. Shilson, Vol. x, page 333; also, Proceedings, I.Mech.E., 'Variable-speed Gears for Motor Road vehicles,' by R. E. Phillips, 1915, page 753.

Nevertheless, the low figure of 1*s* 5*d*. per mile given in Colonel Crompton's original Paper compared quite favourably with present day expenditure on the tyres of public service vehicles. The evolution of the girdle for haulage vehicles from the Boydell form shown in Fig. 38 (page 24), was an example of the resuscitation, under better conditions, of an idea which had been abandoned for a time.

Mr. Watson had raised the question of the use of steam for chain-track tractors. This was done actually in the Log-Hauler, Figs. 59 to 67, Plates 9 to 11, and in the Model 60 Parsons Ditcher.

There was no reason why the chain-track tractor should not be run by steam if so desired, but it should be remembered that the steam-tractor required some extra labour for firing and also time for lighting up, and that it required a much larger supply of water than the total fuel and water capacity provided for the internal-combustion engine.

The picking up of stones and dirt by the chain-tracks was a difficulty with which all the makers were contending, by designing the chain-links or feet so that the joints were better covered and protected.

A great point in designing chain-tracks was the realization of the fact that they were actually geared to the ground surface; increase in the depth of immersion when running on soft ground increased the area of ground to be sheared, and provided more resistance within certain limits, Fig. 95 (page 49) showed how far the track must be pressed into soft ground to get full immersion of the chain; this might vary from a depth of 2 inches to 3 inches according to the make of chain-track used. The figures for the area of tracks given in Table 2 (pages 78 - 83) were all calculated on the projected area of the chain over so much of its length as would be covered by the depth of full immersion; that is to say, the chain-track was supposed to be run on ground so soft that it would just depress the soil to the depth of the chain as shown in side elevation, and that the level taken would run out on the sloping portions of the ends. The Author was glad of the opportunity of making this statement which defined the data for the insistent weights as given on page 81 of Table 2.

With regard to the question of speed, the American speed for ploughing was generally given as 2 ½ miles per hour. The Author could not give the full conditions from personal experience, but he could quote that with a machine of 45 h.p., such as shown in some of the illustrations, the amount ploughed was about 20 acres per day. This speed was nearly treble the record obtained in recent returns for the southern English counties now being ploughed by wheeled tractors. The increased rate was obtained by increasing the number of plough-shares rather than by increasing the actual speed of ploughing.

Major Beaumont was perfectly correct in saying that the necessity for the third differential was not entirely eliminated by either the M. & S. or the Walter differential; although the Worm differentials did not absolutely dispense with the necessity of a differential between the front and back axles, they performed their intended function in turning curves. It was perfectly true that extra work was required of the differentials on axles which comprised steering joints, because the articulations or Hooke's joints on each end of the axle were not kept in phase as was the case with the similar joints on the Cardan shaft or propeller-shaft of the ordinary car.

The electric drive was, of course, quite easy to control by means of a very simple controller, and, as would be seen in Fig. 32, Plate 6, the amount of complication introduced by separate electric motors in each wheel was that there were in all eight bevel-pinions and eight bevel-wheels with the independent electric drive, whereas with the mechanical arrangement of three differentials there were four bevel-wheels as well as the whole of the gear of three differentials amounting to twelve more bevel or mitre wheels.

Communications.

Mr. ARNOLD A. ARNOLD wrote that the references in the Paper to the insistent load for chain-track tractors, also the means of steering when this was by the tracks alone, pointed to the necessity of providing such vehicles, when intended for use on soft ground from which a man would be unable to extricate himself if he stepped off the machine, with at least three large chain-tracks in case of one becoming broken in service; otherwise, on account of repairs to the broken track being extremely difficult, if not impossible, due to the nature of the ground on which they would require to be effected, the tractor could not even be hauled out readily, because it would probably be unstable, and the insistent load would be very greatly increased. The case would also be further aggravated as the tractor would have no means of proceeding by its own efforts, because the means of propulsion were then impaired and the means of steering entirely absent; it could then only take a more or less circular path or would turn round continuously. That is to say, the breakage of a single chain-track at once robbed the class of tractor referred to of means both for driving and for steering, which was a very serious matter under the circumstances stated, and placed this type of tractor at a very great disadvantage as compared with those having means of driving independent of those for steering; a tractor of the latter type with one chain-track broken might possibly proceed, although somewhat erratically in its course, over open country.

With three chain-tracks, should the central one fail, the insistent load would be somewhat increased, and the steering would not be materially affected. Should an outer chain-track fail, the tractor with a wide central chain-track to give a suitably low insistent load, might proceed, if not unstable, with the two remaining chain-tracks by means of which steering could be effected although somewhat imperfectly. Facilities for separate operation of the three chain-tracks would be essential, and the construction might therefore be complicated, but might be warranted by circumstances.

It was not quite apparent why the weight of engines was of little importance, and why the engines should run at comparatively low piston-speeds, thereby increasing the weight for a given engine power. Unnecessary weight in the engines increased the loads on the chain-tracks and placed the centre of gravity of the tractor higher than would otherwise be the case, both of which effects would appear to be disadvantageous, especially for service on bad roads. As a large speed-reduction would probably be necessary, the provision of gearing for that purpose should not be insurmountable, except perhaps on the score of expense.

Mr. VICTOR F. FEENY wrote that the chain-track vehicle was at present, he believed, an illegal vehicle for use on roads, in that it did not comply with the Act regulating motor traffic on highways. It was to be hoped, in view of the immense future before this type of vehicle, that the Local Government Board would see that the necessary legislation was passed.

A similar law was still in force in California, U.S.A., but some years ago it was decided by the State Government that chain-track vehicles would be allowed on all State Highways without fitting any cushioning attachment to the track shoes. The California State Highways were about the finest in the United States, and the top dressing there was soft asphalt. It was remarkable how little the good roads were marked or damaged by chain-tracks when grousers or spuds were not fitted.

The best published acreage per wheeled tractor belonging to the Food Production Department in the United Kingdom was about 35 acres per week. The fuel consumption was also about 5 gallons per acre. With an efficient chain-track tractor of equal horse-power, there would be no difficulty in averaging sixty or more acres per week, because difficulties which prevented a wheel-tractor operating on soft land were avoided in the case of a chain-track tractor. The fuel consumption per acre would also automatically decrease, because of the difference between the insistent loads of the wheel and chain-track tractors.

A tremendous impetus had been given to the chain-track movement in the United States, and the figures relating to the chain-track tractors (all of one make) ordered by the American Government were surprising. For hauling guns the wheel-tractor had been entirely eliminated. It was much to be regretted, in view of the desire to put so many more millions of acres under cultivation in the United Kingdom, that more chain-track vehicles could not be obtained, but military requirements seemed to have taken precedence entirely.

Mr. HENRY MCLAREN wrote that Mr. Legros, when in Leeds, gave them a very interesting Lecture. As Chairman of the Leeds Meeting, he not only thanked Mr. Legros and Mr. Amos, but also the Council of the Institution for having this Paper read in Leeds, where many members and others were employed in making this class of machinery. He proposed to confine his remarks to the ploughing section of the Paper contributed by Mr. Amos.

Before serving his apprenticeship to engineering, he (Mr. McLaren) was intimately connected with farming and the practical working of steam-ploughs on the farm. Forty-two years ago, in conjunction with his brother, they established their present business for the manufacture of traction-engines and steam-ploughing tackle, and to-day their output was about equally divided between these two branches. Therefore both sections of the Paper interested him. His remarks might be unfavourable to the motor-tractors for use in this country as a commercial proposition, yet he was quite in accord with the Government policy of using them as much as possible during the war, especially at the moderate charge to the farmer of 15*s* per acre, even if the actual cost were three or four times that amount.

Mr. Amos, stated quite rightly, that ploughing done immediately after harvest was best, and gave good and sufficient reasons for his contention; but when he stated that the chief value of the motor to the farmer was that it enabled him to get this done, he (Mr. McLaren) replied that the farmer did not get it done quickly enough. Recently Mr. George Lambert, in the House of Commons, asked for a return showing the number of Government tractors in use and the acreage ploughed. Mr. Prothero, in his reply, gave a Table of Results, Table 3 (page 95), and also stated that the weather had been exceptionally bad. He had therefore added a fourth column to the Table, showing the percentage of time lost through bad weather, and he (Mr. McLaren) had added a fifth column, showing the acres that would have been ploughed had there been no time lost by bad weather. It would be seen that, during the six weeks ending 12th January, the average acreage ploughed per week per tractor was only 4.39 acres per week, or 0.73 acre per day. The fifth column showed that had there been no lost time through weather, the average would have been 6.56 acres per week per tractor. Therefore the work done by each tractor was just about equal to that expected from a single-furrow plough, worked by a ploughman and a good pair of horses. These results were altogether too meagre.

Tractors depended upon light lands for getting up their averages, (Mr. Henry McLaren.) yet he had heard of a case in the North of England, where a Government tractor, and a man with three horses and a double-furrow plough, started work at the same time on easy going land. When the tractor had finished, the lands were measured, and the result was in favour of the horses in the proportion of 8 to 5. There was no breakdown of the tractor, but just the usual frequent stops for adjustment, which proved so irritating to the users of these machines.

TABLE 3.

Week ending.	Number of Tractors in commission in hands of County Committees.	Number of Acres ploughed.	Average Number of Acres ploughed per Tractor.	Working hours idle on account of weather.	Working full time.
				Percent.	
8th December 1917	1,708	12,609	7.38	11.13	8.3
15th December 1917	1,721	12,417	7.22	13.36	8.33
22nd December 1917	1,718	5,020	2.92	52.71	6.17
29th December 1917	1,733	2,090	1.21	67.33	3.7
5th January 1918	1,760	8,303	4.72	30.72	6.8
12th January 1918	1,813	5,187	2.86	52.85	6.06
Average for 6 weeks	-	-	4.39	38.0	6.56

Owing to the shortness of food, everyone was interested in the progress of the ploughing tractors. A few days ago it was stated in a Leeds newspaper that 27 acres had been ploughed in one week by a single motor-tractor, and that a week's favourable weather had enabled 200 tractors in Yorkshire to plough 1,300 acres. This was an average of 6.5 acres per week per tractor, which confirmed the figures in the fifth column of Table 3 for favourable weather It also mentioned that the majority of the Yorkshire conditions. tractors weighed 2 tons and pulled three-furrow ploughs, but that lighter tractors, pulling two-furrows, might be preferable for Yorkshire. This opinion might be in anticipation of the six thousand two-furrow Ford tractors which were now due.

He agreed with the remarks of Mr. Amos on the effect of heavy engines passing over certain lands in unfavourable weather. He remembered a field of grain, ripe for harvesting, showing two green trails where the previous year a ploughing engine had passed over the ploughed land to get out of the field. Not only were the wheel marks green, but the heads of grain were stunted, and of poor quality. Yet he had known the same engines pass over that same field many times without leaving any trace of damage on the following crops. It all depended on the condition of the land, and that in turn depended on the, weather. He was now referring to clay lands; there were other lands where not a trace of damage could be found under any circumstances.

He would now deal with the question of heavy versus light tractors for use in Britain. If the British farmer only used his tractor for ploughing immediately after harvest, pulling as many ploughs as possible for so long as the good weather lasted, he would do *more* and *better* work with heavy tractors than was possible with the light machines working all through the winter months. For winter ploughing in this country the light tractor was a necessity, and the lighter the better.

His (Mr. McLaren's) firm only made steam-tractors. The lightest weighed 5 tons, which they did not recommend except in special cases. There was a wide zone on the earth's surface peculiarly adapted for tractor ploughing, where troubles incidental to the British climate were eliminated, and they confined their efforts to this zone. His firm had been competitors at most of the important European Tractor Ploughing Competitions, including the Royal Agricultural Society's Trials in England, and had been uniformly successful in gaining the First Prize. This was partly due to the fact that these competitions were usually held immediately after harvest, before the land had become sodden by rain. They generally used an 8-ton tractor in these European Competitions, and ploughed from 1 ½ to 2 ½ acres per hour.

Mr. Amos mentioned the question of narrow headlands. His own practice was to leave a margin at each side of the field, of equal width to the required headlands, and then to plough all round. It made no difference to the crop whether the land was ploughed north, east, south or west, but it made a considerable difference to the amount of land ploughed, when it was possible to plough all the time, instead of running empty along the headlands. Mr. Amos stated there was no motor-drawn plough that could deal with hard baked sun-dried clay land. His own firm was dealing with this class of land regularly. About eight years ago in India they ploughed 521 acres in forty days - an average of 13 acres per day - 8 inches deep, at a cost of 4*s.* 9*d.* per acre, and made a first class job, with a tractor weighing 13 tons pulling extra strong ploughs specially made for the job. The land was absolutely dry and terribly hard, and the ploughs chattered exactly like a road scarifier, so they had to rivet up every bolt in the ploughs to keep the nuts in place, and fit the share-bolts with spring washers, and even these needed constant attention. Since then he had seen land in Central Africa, equally as hard as the Indian land, but without a semblance of a crack on its surface. Up till recently this land could only be ploughed when moist; then immediately it was ploughed the vegetation again sprang up, so that before seed time it had to be ploughed again. They were now able to let that land get thoroughly dry, and then to plough it with special direct traction ploughs. There was not a green blade growing on it until the rains came.

He considered the double-engine cable system to be the only suitable method of mechanical ploughing in this erratic climate. The cable system was the survival of the fittest. It had carried on through all the bad times of agriculture, contractors ploughing for as little as 7*s.* per acre. The engines and tackle were well made, of the best materials, and appeared to be everlasting. There were numbers of these sets working in England to-day, which were over forty years old, whereas the life of a tractor was about four years. Double-engine tackles were now being used by Government, and were useful for contract work.

Mr. ROBERT E. PHILLIPS wrote that he would like to call attention to the fact that Mr. Amos was not alone in advocating the use of a balanced or duplex one-way plough driven by a self-contained motor unit, for as far back as 1895 a patent was granted to one L. B. Bethell, of Kingswood Warren, Epsom, for a balanced one-way plough having an internal-combustion engine power unit centrally arranged. He agreed with Mr. Amos that with a machine of this type the width of the headlands and the time occupied in turning could both be reduced to a minimum. The drawback, however, to this type of machine was that it could not be adapted for any other use, that is to say, so readily adapted that its conversion would be a paying proposition.

With an ordinary tractor multiple-plough, the width of the headland to be subsequently ploughed could be materially reduced by employing a form of lifting tackle which, when lifting the multiple-plough would raise the front plough in advance of the rear plough or ploughs, and when lowering the same, would lower the front plough in advance of the rear plough or ploughs. This enabled the furrows turned by all the ploughs to be continued right up to the same, line at the headlands, and in effect reduced the width of the headland to be subsequently ploughed to the overall length of the plough itself.

The most practical solution of the one-way plough which would work with a minimum width of headland, that had come within the writer's notice, was one in which the two ploughs were arranged side by side, and were so mounted that they could be swung transversely in relation to the frame of, the tractor to bring them into and out of their operative positions after having been raised clear of the ground by the lifting mechanism before referred to. This arrangement enabled a very compact machine to be, produced. The overall length was reduced by at least the length of one plough, and the overall width was not increased. The experimental machines which had been built on these lines gave most hopeful results.

Mr. Amos had made no reference to one great defect in the ordinary tractor-plough which the one-way plough overcame, that was the tendency for the plough to slip away from its work when ploughing across a slope and the furrow slices were being turned in the direction of the up gradient of the slope unless it was directly controlled by an operator following the plough, which called for the employment of an additional man who must, moreover, be more or less skilled.

Mr. Amos referred to the tendency of tractors to slide down hill when travelling across a slope, and the damage done thereby. This was without doubt a difficulty that had to be overcome, and the writer thought that this end would 'be attained by employing a steering of the centrally articulated type which lent itself particularly well to agricultural tractors, as the motor unit could be carried by the front part of the frame and the tilling implement by the rear part thereof. If, in conjunction with such a method of steering, a single rear-wheel capable of turning on a vertical axis were also employed, the tractor could be turned in a minimum space relative to the overall; length of the machine.

He thought the Institution was greatly indebted to Mr. Legros for his very comprehensive Paper on a subject which was much to the fore at the present time: The Paper was one which did not call for criticism, and the only remarks he offered were entirely from the historical point of view. In the first place he (Mr. Phillips) invented in about 1882 the double-driving gear referred to as the 'Hedgeland.' This mechanism was nothing more than a double driving two-way clutch, and was designed for transmitting power equally to the two driving wheels of a tricycle without using the 'Starley' balance gear.

In the next place, the credit for the 'positive drive' or limited action - as he preferred to call it - differential gear. was due to Gavin C. Goodhart and the 'M. & S.', the 'Walter' and the 'Wolseley' gears described by the Author were based on the principle enunciated by Goodhart, namely, the introduction into the gearing of a pre-determined amount of Internal friction in proportion to the driving torque. In Mr. Goodhart's experimental gears, some of which were made as far back as 1912, the internal friction was produced either by increasing the friction created by the end-thrust of the wheels of the gear or by increasing the friction between the teeth of wheels. From recent trials of heavy motor tractors with solid axles, that is, axles in which both driving wheels were fixed solid with the axle - no differential being used - it would seem that even for road work the differential gear could be dispensed with, without detriment either to the tyres or to the vehicle generally. If this were so, it would seem to point to the fact that the differential gear could be dispensed with on agricultural tractors and other vehicles intended to travel principally on the land.

Mr. ROBERT ROYDS wrote that, in his opinion, the title of Mr. Legros' Paper seemed somewhat more comprehensive than the treatment of the subject warranted, and, judging from the illustrations in the Paper, there appeared to be too much bias in favour of the internal-combustion engine. Whilst it was probably true that the oil-engine was well adapted for American conditions of oil supply, the general problem could not be separated from the motive-power engine, and it was somewhat doubtful whether the oil-engine was to be preferred to the steam-engine tractor for places such as the British Isles, India, South Africa, Australia, South America, etc. To make the Paper more in keeping with its title, the writer would suggest that it would be a decided advantage to manufacturers of tractors in this country if Mr. Legros could see his way to summarize the relative position of steam-tractors and oil-engine tractors as regards adaptability, first cost, cost of upkeep, cost of fuel, etc.

Mr. WILFRID L. SPENCE wrote that he could recall no recent Paper on automobile engineering matters which had interested him quite so much as this one on what; might almost still freely be described as 'Freak Tractors.' The Author's presentation was timely, and his very full treatment was no more than commensurate with the importance of a subject whose rapid development and the diversity of applications of which came as a revelation. The writer's study of Mr. Legros' Paper, together with his own observations, and a careful perusal of land tractor reports and correspondence had led him, for his own guidance, to formulate the following propositions and deductions:-

It was quite wrong to affirm (as had constantly been the case) that when a two-wheel drive 3-wheel or 4-wheel tractor failed to pull, the ground was necessarily unfit for cultivation. There were in fact in our climate very many days when the land, too soft to carry the weight or too slippery for the grip of such tractors, would be ploughed or cultivated by horses or by a rolling-track tractor.

Consequently no land tractor should, other things being equal, weigh one pound more than was absolutely essential to secure adequate tractive adhesion for its maximum duty. Steering wheels and steering rolling-tracks not driven, and of course, mere load-carrying wheels, contributed nothing to adhesion, they were evidence of defective design and should be eliminated. Arising from the foregoing, the entire weight of a completely successful land tractor must be carried on the driving wheels or rolling-track. It was not an effective argument against this to say that all the tractor weight was, under heavy tractive effort, transferred to the driving wheels, for in that case not only would steering be impossible, but the intensity of loading on the driving wheels would be destructive to the land, There remained therefore three broad types only which could be fully successful under bad conditions, namely:-

(1) The two-wheel tractor, both wheels driven, for light work.
(2) The four-wheel drive tractor.
(3) The pure chain or rolling-track tractor (without steering wheels or steering rolling-track).

The chain-track, other things being equal, added nothing to adhesion - its purpose was to prevent sinking in excessively soft ground by means of increased contact area and reduction of load intensity, hence it was not superior in effective tractive effort to a four-wheel drive tractor. If the four-wheel drive. tractor could be built with wheels large enough to obviate sinking in soft ground, it was obviously superior to the rolling-track machine in respect of simplicity, first cost, maintenance, and transmission efficiency. There should be no insuperable or even great difficulty in designing four-wheel drive tractor to meet the conditions named. Such a machine would do everything that a chain-track machine could do except, perhaps, climbing steps of certain pitches which may demand actual 'caterpillar' or 'tank' action, the gradual fore-to-aft transference of the ground contact. Such a condition might be exemplified by a small hog-back bridge of high rise leaving both fore and aft wheels in the air, For ultimate simplicity, low capital cost, low maintenance and high economy, this four-wheel drive tractor should have no differential gears, no exposed gears of any kind whatsoever, no universal joints and no steering gear. It would be steered by power like the pure chain-track machine, by clutch action or the equivalent.

For positively marshy ground where a loading of the order of 5 lb. per square inch or less was necessary, it was thought that a flexible band rolling track could be added over the driving-wheels. This band would be straked for grip and supported by closely set rollers as for a chain track. Built as a double-ended machine, running indifferently in both directions and carrying reversed ploughs at each end, as with steam-ploughing tackle, this equipment would save all the time wasted at headlands, and with three 12 inch ploughs in a sufficiently large field might easily plough 10 acres per day of ten hours.

Mr. L. A. LEGROS wrote that he had been glad to learn from Sir W. A. Tritton that the description of the 'Centipede' quoted in the footnote on page 36 was inaccurate, and the explanation was that his firm, Wm. Foster and Co., Ltd., were at the time building two types of machines, and the description referred to had become mixed and contained details of both, hence the apparent inefficiency alluded to by the Author.

Mr. A. A. Arnold had mentioned (page 93) the need for three chain-tracks for vehicles intended for use on very soft ground. This question had been considered for other than agricultural conditions, but for quite a different reason, which did not arise in agriculture. If the ground were so soft that a man could not walk over it, and therefore that the vehicle could not be drawn out, those conditions that might cause fracture of the chain-track were not likely to be found.

The case might be taken as parallel to the driving over ordinary roads of a four-wheeled vehicle which would be disabled by fracture of either an axle or a wheel; nevertheless the six-wheeled road vehicle had found but few advocates in the past and had not, any vogue at present.

The reason why the weight of the motor was of little importance was to be found in the fact that the weight of four-cylinder commercial-vehicle engines, with cast-iron crank-cases, was about 22 lb. per h.p., whereas-the weight of four-cylinder passenger-car engines was about 11 lb. per h.p, The saving in weight for the engine of a 50 h.p. chain-track tractor would, accordingly be some 5 cwt., and as this would only amount to about 4 per cent. of the weight of the vehicle, it would be consequently hardly worth considering. The use of slow-speed engines running at about one-half the piston speed of the passenger-car engine contributed to longer life of the power-unit, and did not demand so large a reduction ratio in the gear, but as the strength and dimensions of the gear teeth were decided by the speed of the vehicle and the power to be transmitted, the ratio of reduction had very little bearing on the weight of the vehicle.

Mr. V. F. Feeny had raised a very important point with regard to the use of chain-tracks on ordinary roads (page 94). It had been pointed out by Colonel Crompton that much damage had been done to roads by hauling heavy loads on vehicles fitted with small wheels, and therefore having a very high insistent weight. A bad case cited by him was that of the ordinary boiler lorry, and he had stated that a scrapped Lancashire boiler hauled on a wheeled lorry by a wheeled tractor from its place to the scrap merchant's yard might easily cause damage to the road exceeding the value of the scrap itself, let alone the value of the scrap less the cost of haulage. This was a very serious consideration, and one that was worthy of the attention both of the Road Board and of the Local Government Board, for the good of the community.

It was highly desirable that the chain-track system should be recognized by those bodies, and that reasonable regulations should be framed for width, form of track, and permissible speeds under the different conditions of running light, as in a haulage wagon, and of hauling a load, as in a tractor, as well as for the maximum insistent loads permissible on chain-track haulage wagons and chain-track tractors.

Since Mr. Feeny's communication was written it had been claimed† that the record for a week's ploughing with a Government tractor had been made by a 'Titan' wheeled tractor hauling a Ransome three-furrow plough, and that 51 acres had been ploughed in 66.5 hours at Redhill (Surrey); the previous record was stated to be 48 acres under long grass, and the fuel consumption was given as 4 gallons of paraffin per acre. Unfortunately in neither case was the depth given, and this was a most important factor in comparing area ploughed, fuel consumption, and cost.

After delivering his Lecture at Glasgow, the Author had been informed that, for producing a satisfactory potato crop from the fields of the Scottish lowlands, a depth of ploughing of 10.5 inches was necessary. Now it was probable that the power required for ploughing varied as the, area of the ground section turned over, that is as the product of by width, or if these were in the same ratio it varied as the square of the depth. Hence it would follow that the expenditure of power per acre would vary as the depth, and to plough 10.5 inches deep would require an expenditure of 75 per cent. more fuel than for furrows 6 inches deep; moreover, the tractor would only be capable of hauling a proportionately reduced number of ploughshares, in this case four instead of the seven it might haul when ploughing only 6 inches deep.

The actual draw-bar pull for ploughing 6 inches deep by 9 inches wide had been measured by Mr. Alan E. L. Chorlton, C.B.E. (Member), as varying from 500 lb. per share at 2 miles per hour in medium soil to 766 lb. per share at 1.57 miles per hour in stiff clay land.* The speed of ploughing in the United States appeared to be about 2.5 miles per hour, and it was claimed that a No. 18 Yuba tractor (35 h.p.) ploughed 20 acres per day to a depth of 6 inches by 14 inches wide at a cost of 40 cents per acre.

For 370 acres the fuel consumption was given as 500 gallons of gasoline, 30 gallons of cylinder oil and 35 gallons of heavy oil, which at the recent price of 3s. per gallon for petrol and 2s. for cylinder oil would be about 5s. per acre for fuel in Britain.

In another instance a tractor of the same power was quoted as doing 18 acres per day of 10 hours with an average expenditure of 18 h.p,, and using 2 gallons of distillate** per acre.

In Wisconsin a Killen-Strait tractor was stated to have prepared 7 to 8 acres per day hauling two 24-inch breakers, and to haul four 14-inch breakers, when dealing with swamp.

† *The Engineer*, Vol. 125, March 1918, page 207.
* Proceedings, Inst. of Automobile Engineers, 'An Agricultural Power Unit,' Vol. xii (1917 - 18).
** See footnote, page 39.

The Holt Caterpillar figures obtained in comparison with horse haulage were as follows:- horse-labour 4.9 to 3 hours per day per horse at a cost of 7.7 to 14.9 cents per hour; or at 1 acre per day for two or three horses respectively, from $0.75 to $1.34 per acre. The Caterpillar tractor ploughed 3,920 acres in 114 days, or 34.3 acres per day, at a cost of 2405 cents per acre. This amount was composed of:- engineer's wages 12.2 cents, distillate 6.1 cents, plough tender 2.7 cents and lubricating oil, extras, etc., 3.5 cents. For harrowing by 45 h.p. tractor the amount was given as 125 acres per day. In all these instances it must be remembered that the fields were of very large size, and there was not the same need for reducing the headlands as in Britain.

At the Manchester lecture the President had referred to the importance of cost in considering the work done by chain-track-tractors, and the Author quoted that an Austin ditcher was reported to have cut over 9.5 miles of ditch, 4 feet 6 inches deep, shifting 50,000 cubic yards of dirt at a cost of only 3 cents per cubic yard.

At the Leeds lecture Mr. McLaren had given the Government charge for tractor ploughing as 15s. per acre, whereas at Glasgow he learnt it was 25s. per acre or double the pre-war price of 12s. 6d. per acre. Seven acres ploughed by one ploughman with two horses for 6 days was apparently considered a fair average in Scotland. The subject of ploughing by tractor had aroused much interest in the north, and demonstrations had been made of twenty-nine tractors at Edinburgh, Glasgow and Perth.‡

The figures given by Mr. McLaren for cost of ploughing (page 94) required some further comment. The charge of 15s. per acre was in force in England and Wales till October 1917, when it was raised to 20s. per acre for land of ordinary quality and for furrow depths of 6 inches in grass and 7 inches in stubble or fallow land, 2s. 6d. extra per acre being charged for each additional inch of depth, with a further extra charge for fields of very small size or irregular shape, or for exceptional conditions.

Where shallow ploughing was found advisable in the interest of food production the charge might be reduced to 17s. 6d. per acre for ploughing light soil to a depth of 5 inches, and for skimming stronger land to a depth of 4 inches.

The charge for cultivating varied in different counties, but was usually one-third of the charge for ploughing the land, or about 6s. 6d. per acre. For double cultivating cross-ways, the charge in some instances amounted to 20s. per acre. A charge of 25 per cent. extra vas made when, as in some parts of Surrey, a land press was used with the plough. For disk-harrowing the charge was about 4s. per acre. Scotland and Ireland came under other regulations.

In order to permit of comparing these prices at a later date with those of horse traction, it would be of interest to note that the present cost of a horse, cart, and man per week was reckoned as follows:-

	£	s.	d.
Horse, including insurance, interest, keep, and depreciation on £75	2	19	0
Cart, including interest, repairs, and depreciation on £20		6	0
Man, including insurances and holiday allowance.	2	8	9
Office expenses		4	0
Total	£5	17	11

‡ See Highland and Agricultural Society of Scotland, 'Demonstration of Agricultural Tractors,' October 1917, William Blackwood and Sons, Edinburgh and London.

Mr. McLaren had, alluded (page 96) to the fact that the direction of the ploughing made no difference to the crop. It might be thought a heresy by farmers, but he (Mr. Legros) would suggest that the most efficient way of ploughing would be to go round continually in a spiral and finish the remaining approximately triangular corners afterwards; by this means there would be no stoppage for turning, and more than three-fourths of the field would be, completed at a continuous run; furthermore, as the insistent load on the chain-track was so small, the greater part of these corners could be done by lifting the plough over the ploughed portion and making two or three more turns round the field.

Mr. W. D, Collins, of Fulfords, Horsham, had since shown the Author examples of ploughing 14 feet wide with a, gang of ploughs fitted with the automatic lifting and lowering gear mentioned by Mr. R. Phillips (page 97). These were hauled by a 75 h.p. Caterpillar, and the procedure was to mark off 28 feet at each end of the field for turning, and leave 14 feet at each side. After all the centre had been ploughed the tractor was taken across, ploughing each end, and running down the sides with the ploughs raised, after which it was run all round the field.

The corners left untouched amounted to the difference between the area; of a ten yard square and a, ten yard diameter circle, or about 22 square yards only. Mr. Collins stated that he could exceed 2 acres per hour when at work, using the heaviest ploughs intended for steam-tractor haulage, He had also used the Caterpillar for cultivating to a depth of from 15 inches to 18 inches, hauling a Fowler cultivator weighing 2 ½ tons with 11 tynes cultivating 9 feet in width.

Mr. Robert Phillips had mentioned (page 97) his prior invention of the positive-drive differential; apparently this was one of the many cases that occurred of an invention being produced previous to the existence of the conditions necessary for its adoption on a commercial scale; to a smaller extent it resembled in this respect the invention of the chain-track. Dispensing with the differential altogether as suggested by Mr. Phillips might not be very detrimental to the vehicle, but it' might well cause considerable damage to the road .by loosening stones at the surface and thus starting the action which produced pot-holes.

In reply to the remarks of Mr. Robert Royds (page 98), the Author had given particulars of two steam-driven machines - the and the Model 60 Parsons Trench Excavator. These examples sufficed to show that steam-power had been applied to chain-tractors as readily as the petrol or the oil-engine, but it must be remembered that the steam-engine required a supply of fuel, a large quantity of water and constant attention over and above that necessary for running an internal-combustion engine. If an insistent weight of 8 lb. or 9 lb. per square inch on the chain-tracks were admissible, there was no reason why steam chain-track tractors should not be made and work successfully under overseas conditions where petrol and paraffin could not compete with local fuel suitable for steam generating.

At the Glasgow lecture it had been pointed out that the Author had not dealt with the large class of light wheeled agricultural tractors, but had confined himself to the two most recent developments - the four-wheel drive and the chain-track tractor. Several of the problems which arose had been treated by Mr. A. E. L. Chorlton, C.B.E., in his Paper to which reference had already been made.* He gave resistance of a spud as 50 to 75 lb. per square inch for land in which the plough resistance would be 9 lb. per square inch; the rolling resistance of a tractor weighing 3.25 tons with driving wheels 4 feet 6 inches in diameter and 15 inches wide was given as 900 lb. total, or 277 lb. per ton, the wheels being fitted with spuds. He also gave in diagram form the displacement of the wheel loads to which the Author had referred on page 6.

* Proceedings, Inst. of Automobile Engineers. 'An Agricultural Power Unit,' vol, xii, 1917 - 18.

The rolling resistance quoted, amounting to practically 12 per cent. of the weight for the wheeled tractor, did not compare very favourably with the rolling resistance of the chain-track; the Ball-Tread tractor was stated to coast down a gradient of 1 in 33 and a 75 h.p. Holt caterpillar, in use in Sussex, was reported as now being fitted with brakes, because it would coast down gradient 1 in 15.

As the wheeled tractor had so many advocates, that of twenty-nine tractors demonstrated in Scotland, twenty-four were of wheeled types and only 5 chain-tracks, and as the Ministry of Munitions (M.O.M.) tractor was also of the wheeled type, an Appendix was being added (pages 105 to 112) giving the leading particulars of some of these tractors; unfortunately it was not possible to give illustrations of these wheeled tractors at work, though four of the five chain-track vehicles used in the Scottish demonstrations were illustrated in the Paper.

Mr. W. L. Spence had mentioned the difficulty of ploughing and cultivating on soft and slippery ground (page 98); in designing the track-shoes of a chain-track, it was necessary that the track should gear with the ground (page 34); with too small an insistent load, the projections on the shoes would not get sufficient bite on clay or grass land.

Actually the conditions were more favourable for the chain-track, for whereas a wheel only engaged one or two spuds, a chain-track might engage on eight to twenty-four or more gearing surfaces most of which were kept in contact by the loading wheels of the truck-frame; in this matter he disagreed entirely with Mr. Spence. The difficult condition to deal with was that which occurred when the interspace between the projections became filled with clay, so reducing the area of ground in shear to less than one-half and causing slipping to occur. One of the most vital problems was that of how to obtain sufficient penetration for effective gearing with the ground when working on the land, and how to alter or adapt the chain-track so as to be capable of running over the road without damage. The tendency of design of low-powered chain-track tractors was to dispense with steering wheels and to steer by clutches and brakes on the two chain-tracks themselves as suggested by Mr. Spence.

The Author was of the opinion that while it would be quite possible to construct a four-wheel drive tractor such as Mr. Spence had suggested, and to steer it by clutches instead of by steering gear, experience in turning rigid wheel-base vehicles, even on the hard road, showed that serious digging action took place, and even the steam-roller required a pivoting action for its front wheel. Such a tractor as Mr. Spence had described would require a large radius at each side of the wheels to avoid this digging action. The flexible band rolling-track would, the Author feared, be a return, under worse conditions, to the belt form of drive long since abandoned for motor vehicles.

Altogether too much case had been made of the complication of the chain-track and of the track frame. The whole tendency of modern machine-design was to obtain commercial efficiency regardless of multiplication of such parts as could be manufactured in quantity. The early bicycle with its direct drive and plain bearings had given way to the safety bicycle with its ball bearings and chain drive, introducing some hundred or more extra parts in order to obtain greater convenience, safety, and efficiency. When the chain-track had been in use for as long as the safety bicycle, and had thirty years of experience behind it, a similar comparison to that of the types of bicycle could probably be made between wheeled and chain-track tractors.

APPENDIX III.

APPENDIX III.
BATES STEEL MULE.

After the Paper was written the Author received particulars of the Bates Steel Mule which differed in many respects from the chain-track tractors described. It has a single central chain-track for driving and two leading wheels carried in vertically pivoted forked bearings for steering. The original feature of the design is that it is arranged to be driven from the seat of any existing appliance - plough, harrow, cultivator, reaper, etc. - originally intended for horse-haulage. The controls are operated by control column consisting of three co-axial wheels and a lever all arranged on the back ends of three concentric tubes.

The control column is carried on a universal joint at the tractor-end which permits it to be swivelled to any required angle. Steering is effected by the middle wheel and tube; the larger wheel in front of this, that is nearer the tractor, is used to operate the clutch; the small wheel at the rear, that is nearest the driver, is used for gear changing. The lever projecting in front of the forward wheel is for the carburettor control. When ploughing, one of the front wheels runs in the furrow, and it is claimed that under these conditions the machine is self-steering, in other words the steering is of the reversible kind; a special feature is that the front wheel forks are capable of lateral adjustment so as to alter the tread of the tractor.

The chain-track is pivoted about the rear (driving) sprocket axle, and the frame is spring-supported from the track-frame at about the centre of the chain-track. The arrangement of carrying wheels and track-frame resembles that of the Killen-Strait tractor.

The particulars are as follows:-

Engine: horse-power 30; 4 cylinder, 4 cycle, bore 4 inches, stroke 6 inches, speed about 1,000 revolutions per minute, driving through a multiple-disk clutch.
Speeds: 1st, 2.35 miles per hour; 2nd, 3.5 miles per hour; reverse, 2.0 miles per hour.
The water and paraffin tank capacities are not given.

The track supports, apart from the driving sprocket and the leading idler, consist of one pair of large diameter co-axial flanged wheels. The track is 15 inches wide, but the length in contact with the ground and the distribution of weight on the track and wheels are not given; the total weight is 5,600 lb., and the insistent weight is given as 4.5 lb., per square inch. The tractor has an overall length of 11.0 feet and is 8 feet 8 inches wide.

The front (steering) wheels are both 7 inches wide by 30 inches in diameter; the track can be varied from 45 inches to 81 inches to allow for cultivating rows 36 inches to 48 inches wide. The clearance under the frame is abnormally high being 38 inches above the ground; particulars of the pitch of the chain-track are not given; the drawbar pull at the ploughing speed is given as 3,200 lb., which is equivalent to transmitting 20 h.p. through the drawbar, or a mechanical efficiency of 66 per cent. The drawbar pull amounts to 57.1 per cent. of the weight of the tractor.

Special precautions are taken to render the magneto waterproof and to protect the sparking plugs from rain or snow. The carburettor intake is fitted with an air filter, and the water system is stated also to be protected from dust. It is claimed that with a three-furrow plough a man does more than twice as much as he could do by horse-traction in the same time, while for such operations as cultivating and mowing he can do three times as much as with the horses. For use with a binder a longer steering and control column can be substituted for the shorter one. An extension bracket can be fitted at the rear of the tractor to enable the driver to ride on the tractor if this is desired, but this requires two men, whereas the intention of the design is that both machine and implement shall be operated by one man.

TABLE 4 (continued on opposite page).

Wheeled Tractors.

No.	Tractor.	h.p.	Cylinders.			Rated.	Main Clutch.
			No.	Bore.	Stroke.		
	Four-Wheeled			in.	in.	r.p.m.	
1	Alldays General Purpose	25	4	3.94	5.12	1,000	cone
2	Clydesdale	25	4	4.25	5.5	900	multi-disk
3	Denning	20	4	4.25	5.75	900	} integrating disk
4	Denning	16	4	3.5	5.25	1,000	
5	Emerson	30	4	4.5	5.0	800	cone
6	G.W.W.	30	4	4.25	5.75	900	expanding
7	Light All-work	30	4	5.0	6.0	600	disk
8	Mann's Light Steam	25	2	4.0*	7.0	450	throttle
9	Marshall	60	4	7.0	7.0	800	cone
10	Marshall	30	2	7.0	7.0	800	cone
11	Mogul	25	2	7.0	8.0	550	expanding
12	Mogul	20	1	8.5	12.0	400	expanding
13	Mogul	16	1	8.0	12.0	400	contracting
14	M.O.M. (Fordson)	22	4	4.0	5.0	1,000	multi-disk
15	Morse Light Farm	20	4	3.5	5.25	1,200	integrating disk
16	Overtime	24	2	6.0	7.0	400	cone
17	Plow Boy	20	4	3.5	4.25	950	expanding
18	Plow Man (Interstate)	30	4	5.25	5.5	850	expanding
19	Titan	20	2	6.5	8.0	500	expanding
20	Weeks-Dungey	26	4	3.75	5.25	900	disk
	Three-Wheeled.						
21	Allis-Chalmers	18	2	5.25	7.0	720	expanding
22	Peoria	20	4	3.75	5.0	1,000	expanding
23	Farmer Boy	24	4	3.75	5.25	1,100	multi-disk
24	Ivel-Hart	22	2†	5.5	7.0	600	disk
25	Kingsway	16	2	5.75	7.0	750	contracting
26	Samson Sieve-Grip	25	4	4.25	6.75	650	expanding
27	Wallis Junior	30	4	4.25	5.75	900	expanding
28	Whiting-Bull	24	2	5.5	7.0	750	contracting
	Two-Wheeled.						
29	Fowler Motor Plough	14	2	4.0	5.0	1,100	cone
30	Moline Universal	18	2	4.75	6.0	900	cone
31	Wyles Motor Plough	*11*	*1*	*5.0*	*6.0*	*1,100*	*cone*
	Attachment.						
32	Eros (Ford)	(20)‡	(4)	(3.75)	(4.0)	-	-

Approximate and *estimated* figures are shown in italics.

* Compound: l.p. cylinder 6.375 in. bore.
† Two-cycle engine.
‡ The figures in brackets refer to the engine of the Ford Car.

(concluded from opposite page) TABLE 4.

Wheeled Tractors.

Speed m.p.h.			Total Weight	Back Wheels.			Front Wheels.			Wheel Base.	Fuel.	No.
1st.	2nd.	3rd.		No.	Width	Dia.	No.	Width	Dia.			
			lb.		in.	in.						
1.75	2.5	5.0	6,700	2	12	60	2	6	36	87	paraffin	1
2.37	4.0	none	5,200	2	10	60	2	4	46	92	"	2
2.0	to	3.5	4,200	2	10	6	2	5	30	85	"	3
2.0	to	3.5	3,800	2	10	46	2	5	30	95	"	4
1.66	2.32	4.75	5,500	2	12	60	2	6	40	93	"	5
2.5	4.0	none	5,350	2	10	60	2	4	46	90	"	6
1.75	2.4	none	2,450	2	12	48	2	6	32	80	"	7
2.0	to	5.0	10,100	2	20	51	2	8	35	97.5	coke or coke	8
2.0	4.0	none	27,400	2	24	78	2	12	54	126	paraffin	9
2.0	4.0	none	10,100	2	18	78	2	9	54	100	"	10
2.0	4.5	none	10,100	2	12	60	2	6	40	112	"	11
2.0	2.75	none	5,600	2	10	54	2	6	36	90	"	12
2.0	none	none	5,600	2	10	54	2	6	36	90	"	13
1.5	2.75	6.75	2,580	2	12	42	2	5.25	28	63	"	14
1.75	to	3.25	3,600	2	10	46	2	5	30	85	"	15
2.5	none	none	4,800	2	10	52	2	5	36	90	"	16
2.0	3.25	none	4,000	2	10	60	2	5	36	100	"	17
2.0	3.25	none	4,300	2	10	60	2	5	36	100	"	18
2.0	2.75	none	6,850	2	10	54	2	6	36	88	"	19
1.63	2.75	4.5	3,900	2	10	40	2	5	30	60	"	20
2.5	none	none	4,800	2	12	56	1	6	32	96	"	21
2.25	none	none	3,900	2	{ 18 8	60 44 }	1	6	30	100	"	22
2.38	none	none	3,000	2	{ 12 6	50 40 }	1	6	22.5	108	"	23
1.8	3.0	none	6,600	1	26	64	2	6	33	98	"	24
2.75	none	none	3,700	2	10	56	1	5	29	99	"	25
2.0	none	none	5,200	2	18	40	1	15	28	*100*	"	26
2.5	3.8	none	8,000	2	12	48	1	8	30	100	"	27
2.41	none	none	4,850	14 8	{ 14 8	60 40 }	1	6	30	*110*	"	28
1.55	2.22	none	2,500	-	-	-	2	6	44	-	"	29
1.0	to	3.0	2,800	-	-	-	2	8	52	-	petrol	30
1.75	*2.25*	*none*	*2,350*	-	-	-	2	7	*33*	-	"	31
2.5	-	-	1,900§	2	8	38	-	-	-	-	"	32

Approximate and *estimated* figures are shown in italics.

§ Of which the Ford Car accounts for 1,550 lb.
II These tractors took part in the Scottish Demonstration. It will be found that the figures obtained by the Author differ in many instances from those in the Scottish Report.

TABLE 4 (continued on opposite page).

Wheeled Tractors.

No.	Tractor.	h.p.	Cylinders.			Rated.	Main Clutch.
			No.	Bore.	Stroke.		
	Four-Wheeled			in.	in.	r.p.m.	
1	Alldays General Purpose	25	4	3.94	5.12	1,000	cone
2	Clydesdale	25	4	4.25	5.5	900	multi-disk
3	Denning	20	4	4.25	5.75	900	} integrating disk
4	Denning	16	4	3.5	5.25	1,000	
5	Emerson	30	4	4.5	5.0	800	cone
6	G.W.W.	30	4	4.25	5.75	900	expanding
7	Light All-work	30	4	5.0	6.0	600	disk
8	Mann's Light Steam	25	2	4.0*	7.0	450	throttle
9	Marshall	60	4	7.0	7.0	800	cone
10	Marshall	30	2	7.0	7.0	800	cone
11	Mogul	25	2	7.0	8.0	550	expanding
12	Mogul	20	1	8.5	12.0	400	expanding
13	Mogul	16	1	8.0	12.0	400	contracting
14	M.O.M. (Fordson)	22	4	4.0	5.0	1,000	multi-disk
15	Morse Light Farm	20	4	3.5	5.25	1,200	integrating disk
16	Overtime	24	2	6.0	7.0	400	cone
17	Plow Boy	20	4	3.5	4.25	950	expanding
18	Plow Man (Interstate)	30	4	5.25	5.5	850	expanding
19	Titan	20	2	6.5	8.0	500	expanding
20	Weeks-Dungey	26	4	3.75	5.25	900	disk
	Three-Wheeled.						
21	Allis-Chalmers	18	2	5.25	7.0	720	expanding
22	Peoria	20	4	3.75	5.0	1,000	expanding
23	Farmer Boy	24	4	3.75	5.25	1,100	multi-disk
24	Ivel-Hart	22	2†	5.5	7.0	600	disk
25	Kingsway	16	2	5.75	7.0	750	contracting
26	Samson Sieve-Grip	25	4	4.25	6.75	650	expanding
27	Wallis Junior	30	4	4.25	5.75	900	expanding
28	Whiting-Bull	24	2	5.5	7.0	750	contracting
	Two-Wheeled.						
29	Fowler Motor Plough	14	2	4.0	5.0	1,100	cone
30	Moline Universal	18	2	4.75	6.0	900	cone
31	Wyles Motor Plough Attachment.	*11*	*1*	*5.0*	*6.0*	*1,100*	*cone*
32	Eros (Ford)	(20)‡	(4)	(3.75)	(4.0)	-	-

Approximate and *estimated* figures are shown in italics.

* Compound: l.p. cylinder 6.375 in. bore.
† Two-cycle engine.
‡ The figures in brackets refer to the engine of the Ford Car.

(concluded from opposite page) TABLE 4.

Wheeled Tractors.

Speed m.p.h.			Total Weight	Back Wheels.			Front Wheels.			Wheel Base.	Fuel.	No.
1st.	2nd.	3rd.		No.	Width	Dia.	No.	Width	Dia.			
			lb.		in.	in.						
1.75	2.5	5.0	6,700	2	12	60	2	6	36	87	paraffin	1
2.37	4.0	none	5,200	2	10	60	2	4	46	92	"	2
2.0	to	3.5	4,200	2	10	6	2	5	30	85	"	3
2.0	to	3.5	3,800	2	10	46	2	5	30	95	"	4
1.66	2.32	4.75	5,500	2	12	60	2	6	40	93	"	5
2.5	4.0	none	5,350	2	10	60	2	4	46	90	"	6
1.75	2.4	none	2,450	2	12	48	2	6	32	80	"	7
2.0	to	5.0	10,100	2	20	51	2	8	35	97.5	coke or coke	8
2.0	4.0	none	27,400	2	24	78	2	12	54	126	paraffin	9
2.0	4.0	none	10,100	2	18	78	2	9	54	100	"	10
2.0	4.5	none	10,100	2	12	60	2	6	40	112	"	11
2.0	2.75	none	5,600	2	10	54	2	6	36	90	"	12
2.0	none	none	5,600	2	10	54	2	6	36	90	"	13
1.5	2.75	6.75	2,580	2	12	42	2	5.25	28	63	"	14
1.75	to	3.25	3,600	2	10	46	2	5	30	85	"	15
2.5	none	none	4,800	2	10	52	2	5	36	90	"	16
2.0	3.25	none	4,000	2	10	60	2	5	36	100	"	17
2.0	3.25	none	4,300	2	10	60	2	5	36	100	"	18
2.0	2.75	none	6,850	2	10	54	2	6	36	88	"	19
1.63	2.75	4.5	3,900	2	10	40	2	5	30	60	"	20
2.5	none	none	4,800	2	12	56	1	6	32	96	"	21
2.25	none	none	3,900	2	{ 18 / 8	60 / 44 }	1	6	30	100	"	22
2.38	none	none	3,000	2	{ 12 / 6	50 / 40 }	1	6	22.5	108	"	23
1.8	3.0	none	6,600	1	26	64	2	6	33	98	"	24
2.75	none	none	3,700	2	10	56	1	5	29	99	"	25
2.0	none	none	5,200	2	18	40	1	15	28	*100*	"	26
2.5	3.8	none	8,000	2	12	48	1	8	30	100	"	27
2.41	none	none	4,850	14 / 8	{ 14 / 8	60 / 40 }	1	6	30	*110*	"	28
1.55	2.22	none	2,500	-	-	-	2	6	44	-	"	29
1.0	to	3.0	2,800	-	-	-	2	8	52	-	petrol	30
1.75	*2.25*	*none*	*2,350*	-	-	-	2	7	*33*	-	"	31
2.5	-	-	1,900§	2	8	38	-	-	-	-	"	32

Approximate and *estimated* figures are shown in italics.

§ Of which the Ford Car accounts for 1,550 lb.
‖ These tractors took part in the Scottish Demonstration. It will be found that the figures obtained by the Author differ in many instances from those in the Scottish Report.

WHEELED TRACTORS.

The Author has prepared Table 4 from data obtained direct from the manufacturers or agents in the majority of the cases tabulated. The Scottish demonstrations included also the Case and Saunderson-Universal four-wheel tractors, the Chase three-wheel tractor, and the Hodgson attachment for Ford cars. Five chain-track tractors were shown, namely the Bates Steel Mule, Burford-Cleveland, Creeping-Grip, Killen-Strait Model 3, and Martin Cultivator, all of which the Author has described.

At the Lectures given at Manchester, Leeds, Lincoln and Glasgow much interest was expressed relating to the various recent developments in light wheeled tractors for farm work in connexion with food production. These might be divided into four main classes: four-wheeled, three-wheeled, two-wheeled tractors, and tractor attachments for converting ordinary cars to haulage work. The first two groups were adapted for road-haulage, whereas the two-wheeled tractors were intended for use with a plough as a complete unit. There were at present well over fifty tractors of these various types many of which were but little known in Britain. The demonstrations made in Scotland, at Edinburgh, Glasgow, and Perth comprised a representative selection.*

It would be seen from this very imperfect survey that the subject of wheeled agricultural tractors was one worthy of a Paper in itself.

(Mr. L. A. Legros.)

Mr. ARTHUR AMOS, in replying to the criticisms on his Paper, wished to thank the members for the kindly manner in which it had been received. He entirely agreed with Mr. Watson's contentions that, given ordinary hard land, it was advantageous to have two of the tractor-wheels running in the furrow, not only because by this means the diagonal pull was avoided, but also because by this method, if the top surface was slightly greasy and the undersoil dry, a common condition in autumn, the tractor-wheels obtained a better foothold and less damage to texture resulted. He also agreed that it was desirable to consider the possibility of arranging a subsoiling attachment to, or in conjunction with, the plough, and this was of special importance if, as indicated above, two of the tractor-wheels ran in the furrow; in this case it was desirable that the subsoil after pressure by the wheels in the furrow should be loosened by some subsoiling implement immediately after the wheels had passed and before the next furrow was turned into the furrow. It was not necessary to subsoil land frequently, and he suggested that this single subsoil attachment following the wheels would be sufficient for the purpose. He did not agree with Mr. Watson, however, that 2 ½ miles per hour would be the best rate at which tractors should work in all cases; in winter, when the ground was wet, he agreed that too rapid a speed left the furrows lying very irregularly, and rubbish was not so well buried, but for ploughing dry land, it was advantageous to break up the furrow, as with the digging type of plough, because this facilitated subsequent cultivation.

---

* Highland and Agricultural Society of Scotland; Demonstration of Agricultural Tractors and Ploughs, October 1917. Since the Paper, went to press a series of articles on British Agricultural Tractors has appeared in *The Engineer*, Vol. 124, pages 512-3, 533-5, 553-6, 562 and Vol. 125, pages 4-6, and 45-6. The descriptions include the Aveling and Porter, Burrell and Son, Foster, Garrett, Mann, McLaren and Ransomes, Sims and Jefferies, steam-tractors, as well as the Bumsted and Chandler, Crawley Agrimotor, Foster and Ivel internal-combustion engine tractors.

Professor Baker had asked him a question upon the damage done by the pressure of the wheels per square inch on the soil (page 91); he could best explain the damage by saying that when clay soil was worked or pressed, when wet, into a pasty condition (as the brickmaker worked his clay for brickmaking) the soil became impervious to air and water, and crops could not be successfully grown upon it - the texture was spoilt.

It was important to remember that pressure upon the land might damage the texture in two distinct layers of the soil; it might damage the surface soil only, as for example in autumn and early winter when rain might have fallen recently upon dry land, in this case comparatively small pressure per sq. inch would effect the damage, so that in this case the chain-track wheels were no better than others; or the damage might occur to the soil 3 or 4 inches below the surface, as for example in second ploughing or cultivating in spring after a spell of dry east wind, when the surface was dry, but the lower depths more or less sodden. In this case heavy pressure per square inch occasioned very serious damage, but a chain-track tractor might be able to pass over the surface with little damage to the lower depths. In agriculture there was a saying that "a March roll makes an April Fool," which meant that if one rolled clay land wheat when only the surface was dry too early in March, the texture of the lower depths would be damaged, and the wheat would look very yellow in April.

With many of Mr. McLaren's criticisms he cordially agreed, but Considered that he had stated the case much too severely against the tractor by using Mr. Prothero's table of the work accomplished by government tractors in December and January, in criticism of the Author's statement that the motor tractor enabled the farmer to get his ploughing done quickly after harvest. There were many factors which contributed to put the work of the tractors, as shown in this Table, in the worst possible light; in December and January the land surface was rarely if ever in good order for ploughing, consequently, the tractor was handicapped to a very much greater extent than the number of hours lost, and for which Mr. McLaren allowed in his Table. The case was made still worse by the fact that in very many cases the tractors were actually at work when, owing to the condition of the land, they ought to have been idle. At this season in the year the days were short; in September and October, after harvest, there were about four more hours of daylight during which time an enterprising farmer would be able to keep his tractor running. The Government only commenced to obtain delivery of their tractors in quantity during the autumn and had been continually increasing their numbers during the season; they had had to train soldiers from the lower grades as drivers ; consequently, during the period quoted, a large proportion of the drivers were mere novices, and it was a matter of common knowledge that tractors required skilled and intelligent driving in order to accomplish good work.

Again, much ploughing time was lost under the Government scheme by the moving of the tractor and plough from one farm, at which it might have been working, to the next, which might be many miles away; not only was time lost on the road, but adjustments of wheels, etc., had also to be made ; in the case of a privately-owned (by the farmer) tractor such loss of time would be reduced to a minimum. In place of the 6.5 acres per week, which Mr. McLaren suggested (page 95) as the average of what tractor was capable in favourable weather, he would suggest that a good privately-owned tractor, driven by the farmer's son or other sympathetic driver, and working on average dry land after harvest, might be expected to accomplish on the average 20 acres per week with the plough. He fully agreed with Mr. McLaren that one of the most serious defects of the average motor tractor was the very rapid depreciation of the machine, a depreciation which tended to make the cost of tractor ploughing very heavy. When the Author in his original Paper stated that e he knew of no tractor capable of dealing with the heaviest types of sun-baked clay, he should have qualified his statement by inserting the word 'light' before tractor; he agreed that steam tractors of 5 to 6 tons were capable of accomplishing very valuable work on such clay land after harvest so long as the surface. remained hard and dry.

Mr. Phillips had drawn attention to the fact that the Author was not alone in advocating one-way ploughing; he freely admitted it, but desired to emphasize that one-way ploughing was preferable from the farmer's point of view upon most land, and that this opened up one of the most promising opportunities for future development of the motor-tractor.

Plate 1.

FIG. 7. *Dorr-Miller Automatic-locking Differential.*

FIG. 10. *F.W.D. Centre Differential and Locking-levers.*

FIG. 11 *F.W.D. Transmission; Phantom view.*

Plate 2. F.W.D.

FIG. 12. *3-ton Chassis.*

FIG. 13 Side-tip Wagon.

FIG. 14 *Front Axle.*

Plate 3.   F.W.D. *3-ton Truck.*

FIG.15 *Carrying Ore over Sand, Utah*    FIG.16 *On a Sea of Alkali-mud (70 miles from base).*

FIG.17 *Hauling Industrial Railway Cars*

Plate 4.

FIG.18. *F.W.D. End-tip Wagon.*

FIG.23. *Walter Super-quad; inverted view-wheels straight.*

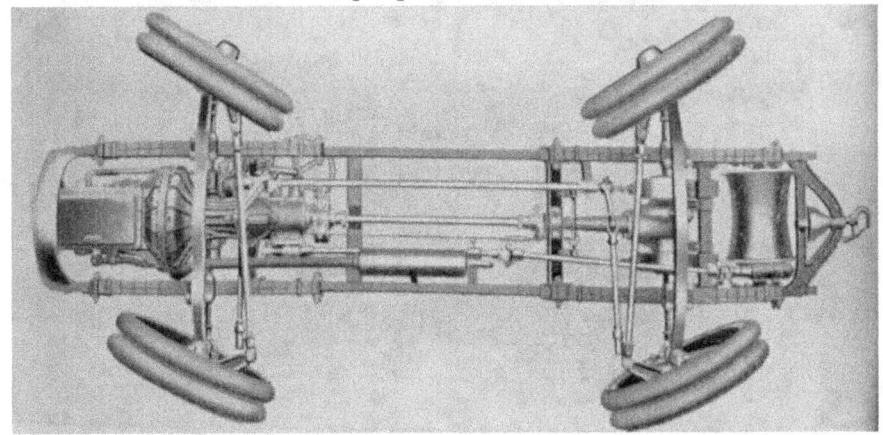
FIG.24. *Walter Super-quad; inverted view-wheels locked.*

Plate 5.

FIG.25. *Walter Automatic-locking Differential.*

FIG.26. *Jeffrey Quad; Tractor Chassis.*

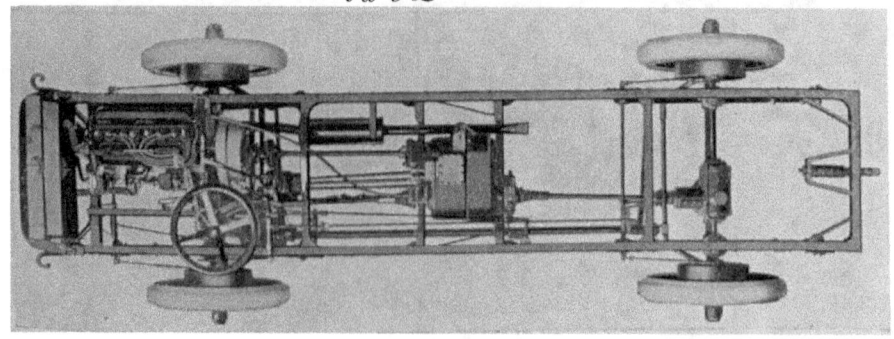

FIG.27. *Jeffrey Quad; plan view of Chassis.*

Plate 6.   *Couple-gear*
FIG. 31. *Wheel, Complete without tyre.*

FIG. 32. *Wheel, details showing Motor.*

FIG. 34 *Petrol-electric Truck with Load.*

Plate 7. *Couple-gear*

FIG.35. *Petrol-electric Truck, Power Plant.*

FIG.36 *Petrol-electric; Fire Escape.*

Plate 6.   *Couple-gear*
FIG. 31. *Wheel, Complete without tyre.*

FIG. 32. *Wheel, details showing Motor.*

FIG. 34 *Petrol-electric Truck with Load.*

Plate 7. *Couple-gear*

FIG.35. *Petrol-electric Truck, Power Plant.*

FIG.36 *Petrol-electric; Fire Escape.*

Plate 8.

FIG.43. *Chain-track of Log-Hauler (Phoenix).*

FIG.44. *Track detail; Modern Pedrail (Diplock).*

FIG.55. *Rocker-pin of Tracklayer (Best).*

FIG.56. *Track-shoe; Pressed-steel.*

FIG.57. *Track-shoe of Steel-casting; one piece.*

Plate 9. *Log-Hauler.*

FIG.59. *With Runners for Steering.*

FIG.60. *Track-detail and Drive, showing Engine.*

FIG.61. *Showing small Ground-clearance for Snow.*

Plate 10. *Log-Hauler.*

FIG.62. Showing relative width of Engine and Sleigh-tracks.

FIG.63. With Load of Sawn Timber.

FIG.64. View of 420-ton Train of Logs.

Plate 11.

FIG.65. *Log-Hauler; on Curve.*

FIG.66. *Centiped Truck.*

FIG.67. *Centiped; Hauling Wagons on Soft Sand.*

Plate 12.   *Centiped.*

FIG. 68. *Crossing Ditch.*

FIG. 69. *Crossing Railway Track.*

FIG. 70. *Detail of Track.*

Plate 13.

FIG.71. *Centiped; Ploughing 30-inch Furrow.*

FIG.72. *Centiped; Road-grading.*

FIG.74. *Caterpillar; 120 h.p.*

Plate 14. Holt Caterpillar.

FIG.75. *75 h.p.*

FIG.76. *45 h.p.*

Plate 15. Catepillar.

FIG.77. *45 h.p., Truck Assembly.*

FIG.78. *18 h.p. with Track laid out.*

FIG.79. *18 h.p., view from above.*

Plate 16. Holt Caterpillar.

FIG. 80. *18 h.p.*

FIG. 81. *Cultivating and Harrowing.*

FIG. 82. *Hauling Ditcher.*

Plate 17. *Caterpillar.*

FIG. 83. Working for White Pass and Yukon Railroad.

FIG. 84. Hauling Logs on Sleigh.

FIG. 85. Clearing Snow at Butte, Montana.

Plate 18. *Clayton.*

FIG. 92. *100 h.p. Tractor*

FIG. 93. *110 h.p. Tractor; Hauling Test Load.*

Plate 19.

FIG. 94. *Clayton 110 h.p. Tractor; on 40° Gradient; front wheel off the ground.*

FIG. 101. *Best Tracklayer; 90 h.p.*

Plate 21. *Creeping-Grip.*

FIG. 107. *'Senior'; 50 h.p.*

FIG. 108. *'Baby'; 16 h.p.*

FIG. 109. *'Baby'; 16 h.p., Cultivating.*

Plate 20. *Best Tracklayer*

FIG. 102. *30 h.p.*

FIG. 103. *Detail of Differential and Drive-sprockets.*

FIG. 104. *Ploughing in Reeds, California.*

Plate 22.

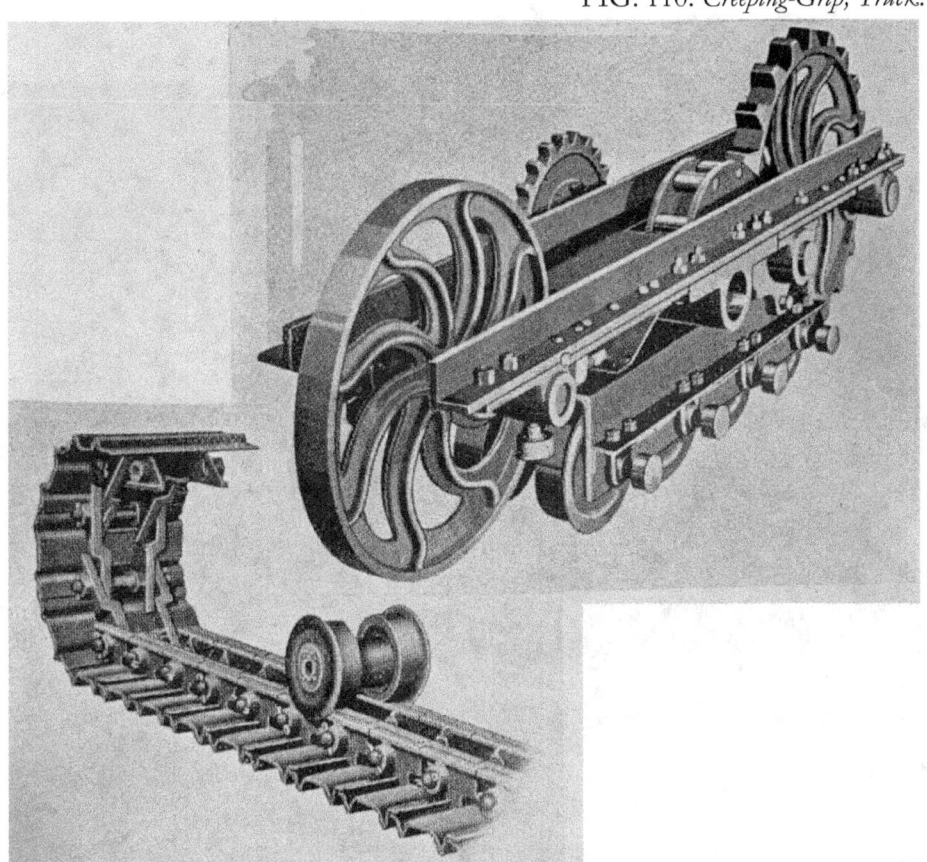

FIG. 110. *Creeping-Grip; Truck.*

FIG. 111. *Creeping-Grip; Track and Wheels.*

FIG. 112. *Austin Tractor; 35 h.p.*

Plate 23.

FIG. 113. *Austin Tractor; 15 h.p.*

FIG. 114. *Burford-Cleveland Tractor.*

Plate 24.

FIG. 115. *Burford-Cleveland Tractor. Hauling Ploughs.*

FIG. 116. *Strait Tractor; Front view, showing Steering-track.*

Plate 25. Strait Tractor.

FIG. 117. *Model 3. Single-track Tractor; Deep Ploughing.*

FIG. 118. *Detail of Track-frame.*

Plate 26. *Strait Tractor.*

FIG. 119. *Side view.*

FIG. 120. *Hauling a House on a dirt road.*

Plate 27. *Strait Tractor.*

FIG. 121. *Ploughing in old sod.*

FIG. 122. *Model 3; Side view.*

Plate 28. *Yuba Tractor.*

FIG. 124. *Side view.*

FIG. 125. *Detail of Truck.*

FIG. 126. *Disking and Harrowing.*

Plate 29. Yuba Tractor.

FIG. 127. *Section of Track.*

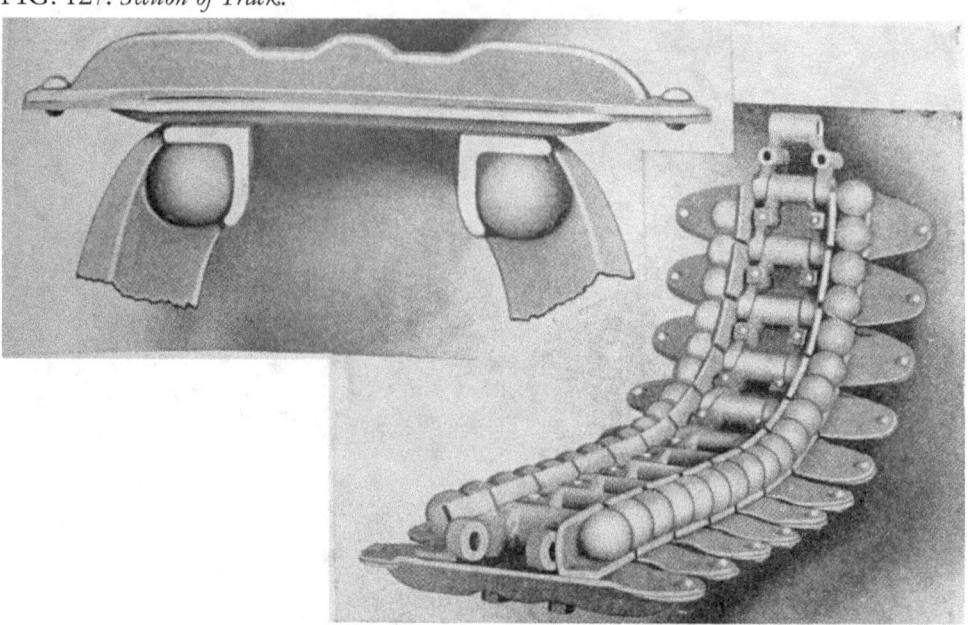

FIG. 128. *Detail of Track.*

FIG. 129. *Detail of Gear-box.*

Plate 28. *Yuba Tractor.*

FIG. 124. *Side view.*

FIG. 125. *Detail of Truck.*

FIG. 126. *Disking and Harrowing.*

Plate 29. *Yuba Tractor.*

FIG. 127. *Section of Track.*

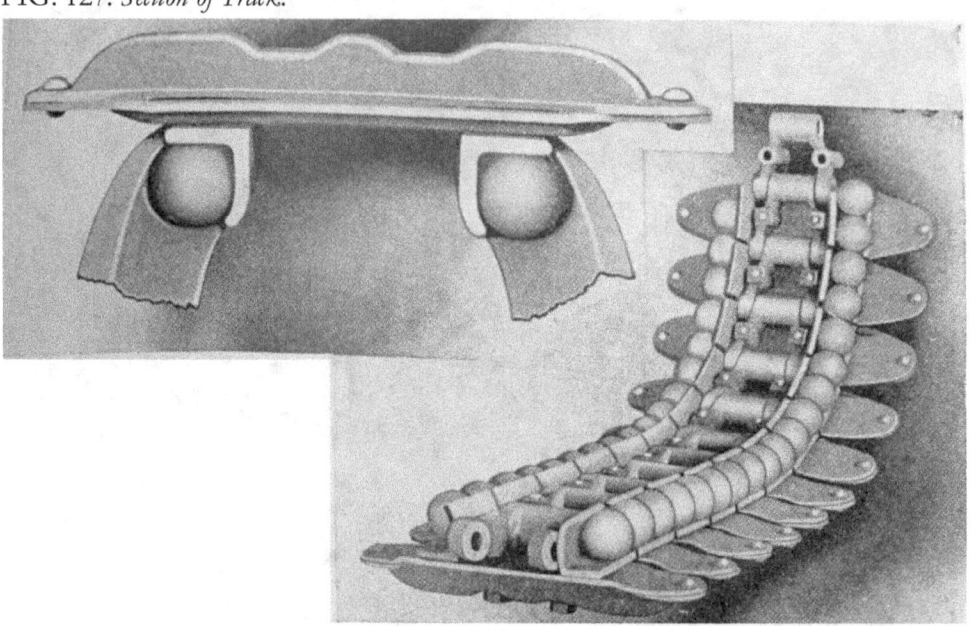

FIG. 128. *Detail of Track.*

FIG. 129. *Detail of Gear-box.*

Plate 30. *Yuba Tractor.*

FIG. 120. *Hauling Harvester.*   FIG. 131. *Hauling 40 ft. Corrugated Roller in Bean-land.*

Plate 31. *Yuba Tractor.*

FIG. 132. *Climbing a 65 per cent gradient.*

FIG. 133. *Fitted with Hand-pan Driller.*

Plate 32. *Yuba Tractor.*

FIG. 134. *On Side-lying ground in Orchard.*

FIG. 135. *Climbing out of Irrigation Ditch.*

FIG. 136. *Harvesting Rice.*

Plate 33. *Martin Cultivator.*

FIG. 137. *Two-Plough.*

FIG. 138. *With Steering Wheels.*

FIG. 139. *With Three-furrow Plough.*

Plate 34.

FIG. 140. *Martin Cultivator; Ploughing.*

FIG. 141. *Martin Cultivator; Hauling.*

FIG. 143. *Scott Motor-Sleigh; on ground.*

FIG. 144. *Scott Motor-Sleigh; on Blocks showing Chain Sag.*

Plate 35.

FIG. 145. *Scott Motor-Sleigh; climbing a slope.*

FIG. 146. *Scott Motor-Sleigh; Test on Lake Fefor, Norway.*

FIG. 147. *Lefebvre Tractor; 45 h.p.*

Plate 36.

FIG. 148. *Holt Caterpillar with Train of Chain-track Haulage Wagons.*

FIG. 149. *Caterpillar with Chain-track Haulage Wagons.*

FIG. 150. *Pedrail Haulage Wagon.*

FIG. 151. *Pedrail Haulage Wagon with Chain-track laid out.*

Plate 37.

FIG. 152. *Parsons' Model 60 Ditcher; Side view.*

FIG. 153. *Parsons' Model 48 Ditcher; Front view.*

Plate 38.

FIG. 154. *Parsons' Model 24 Ditcher; Back view.*

FIG. 155. *Austin Ditcher; Digging 10 feet deep, 27 inches wide.*

Plate 39.

FIG. 156. *Austin Ditcher with Bank-sloping Device; Digging 5 ft. deep, 3 ft. bottom, 1 to 1 slope.*

FIG. 157. *Parsons' Back Filler.*

Plate 40.

FIG. 158. *Renard Road Train, with 20-ton load crossing swampy field.*

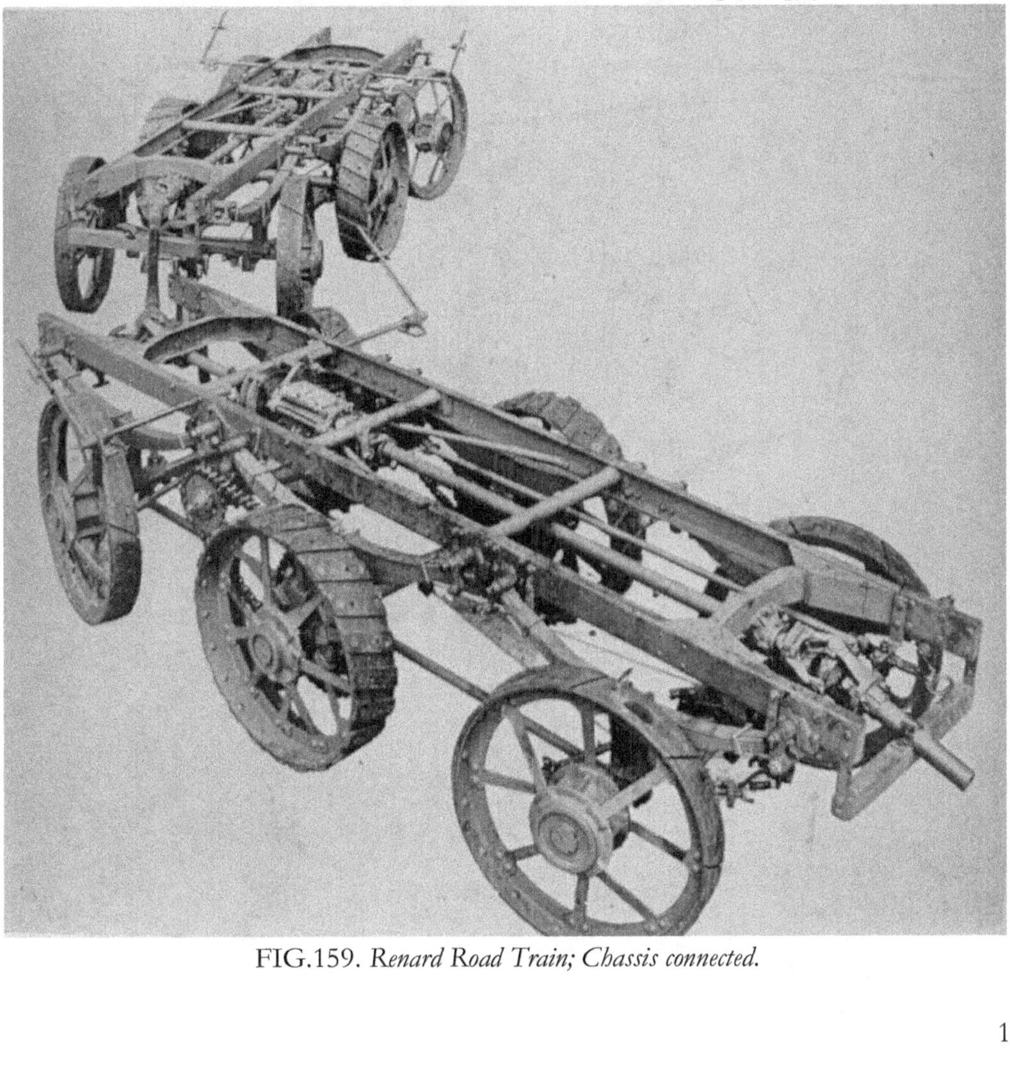

FIG. 159. *Renard Road Train; Chassis connected.*

Tanks and Chain-Track Artillery.

An abstract of paper read before the British Section of the *Société des Ingénieurs Civils de France*. September, 1921 published in 3 parts in 'The Engineer'

    Part I   2nd December 1921
    Part II   9th December 1921
    Part III   16th December 1921‡

‡With a letter of correction from LeGros published 23rd December 1921

A paper published in 2 parts in 'The Automobile Engineer'

    Part I   March 1922

    Part II   April 1922

By L.A. Legros

Tanks and Chain-Track Artillery.*
By L. A. Legros

* Abstract of paper read before the British Section of the Société des Ingénieurs Civils de France. September, 1921.

DURING the last war appliances came into being based on the use of the chain-track as a method of propulsion, which are capable of being divided into two classes:-

(1) The 'tank,' as it is commonly known, an armed and armoured offensive machine, capable of running either on the road, or over uneven ground, trenches, embankments, barbed wire entanglements, &c.
(2) Chain-track artillery, capable of carrying heavy artillery, that is to say, an automotor cannon or howitzer, able to run over and kind of ground behind the firing line.

TRIAL AND EXPERIMENTAL PERIOD.

Methods for crossing the enemy defences were simultaneously subject, of investigation in France and in England at the commencement of 1915, when trench warfare began. In France, as in England, under the seal of secrecy and unknown to each other, engineers were working in adapting the caterpillar to military transport. The first, designs were made in France by Messrs. Schneider et Cie., who had obtained two Holt caterpillars in January 1915, the one having steering wheels and the other of 'Baby' type. These machines were used for tests at Le Creusot in May, 1915. It was found that the 'Baby' type, without, steering wheels, was capable of being handled very readily, and was suitable in principle for an offensive machine capable of running across country and traversing obstacles. Demonstrations were made in the presence of the President of the French Republic, June 16th, 1915. In July, Messrs. Schneider et Cie. Commenced the designs for an armoured chain-track-driven motor vehicle. The apparatus was known as the tracteur armé et blindé.

An order for ten of these machines was given to constructors on December 15th, 1915. The machines were intended to be fitted in front with a wire cutter on the Breton-Prétot system. In the meantime, on the initiative of the deputy, M. Breton, the 'Baby' caterpillar of Messrs. Schneider et Cie. Was used for demonstrations on the uneven ground of the front at Souain, on December 9th 1915, and at Satory camp on December 27th, 1915. Trials were made with an upturned prow and rear wings, which increased its ability to cross obstacles, trenches, parapet, &c.

About the same time, Colonel (now General) Estienne, who had seen the Holt caterpillar at work on the British front, appreciated that the chain-track principle should render it possible to produce land ships which, if used in large numbers in conjunction with infantry, would afford invaluable help in offensive operation. Being placed in communication with the Schneider works on December 10th, he obtained every possible assistance from the firm, as well as being shown the design of the armoured motor machine-gun in a far advanced stage.

A new design was at once put in hand according to the particulars which he supplied, and on December 27th, Colonel Estienne communicated the scheme for the Schneider-Estienne tank to G.H.Q. (G.Q.G.). In order to verify whether the dimensions adopted by Messrs. Schneider properly fulfilled the requirements involved in the crossing of trenches, the Automobile Technical Section made up a temporary machine of the same length of contact of the chain-track by combining the details of two Holt caterpillars.

The trials took place on February 21st, 1916, at the Vincennes Polygon; they showed that the proportions adopted by the constructors were suitable. Consequently, on February 25th, an order was given to Messrs. Schneider for 400 tanks in place of the previous order for the ten motor machine-guns. It is interesting and pleasant to announce that the trials and designs made by Messrs. Schneider were entrusted to our distinguished colleague, Mr. Eugene Brillié.

Furthermore, the roughly constructed machine just mentioned, with tracks assembled from the parts of three Holt chain-tracks, served as a basis for new pattern of tank, the design of which was undertaken by the firm of Saint-Chamond, and resulted in an order for 400 other tanks being placed with that firm. The details of these tanks will be dealt with later. The tanks were in the first instance styled armoured tractors, so as not to attract attention to their object; later they were called 'Chars d'assaut,' the name which was given to them by General Estienne.

While these studies were being pursued in France, Mr. Winston Churchill, Chief Lord of the British Admiralty, formed in England, in February, 1915, a Committee known as the Admiralty Land-ship Committee, of which the consulting engineer was colonel R. E. B. Crompton, C.B., R.E., and the assistant consulting engineer, L. A. Legros. A grant of £80,000 was assured by the British Treasury.

From the commencement Colonel Crompton recognised that it would not be possible to use wheels, of any size whatever, for transport across marshy ground such as that of Flanders, and that the only method of overcoming the difficulty was to be found in the use of chain-tracks. He ordered immediately tractors of the Strait and Creeping-Grip types from America to serve as a basis for experiment, while attempts were being made to adapt the only English chain-track, the Diplock Pedrail, all the available Holt tractors being reserved for the use of the British Army.

It is interesting to note that in most countries, when a Government department is entrusted with work somewhat outside its proper province, it does not receive much help from those which it overlaps. England did not prove to be an exception to this rule, and Colonel Crompton, as Consulting Engineer to the Admiralty, was refused permission to visit the front and to ascertain on the spot the conditions of the ground and of the obstacles to be overcome. It was not till the end of July, 1915, that Colonel (now General) E. D. Swinton, C.B., D.S.O., was able to obtain particulars, semi-officially, on this subject, and to bring the necessary data for the definite design of a land-ship to England.[1]

France, the cradle of the automobile, and the neighbour of Germany, with the large variety of ground in which its frontiers lie, from the dunes of the north through the swamps of Flanders to the mountains of the Vosges, was thoroughly informed on these points and knew the difficulties only too well.

Of the American chain-track tractors available, only two were of possible value in the question of designing the tank in England:-

(1) The 'Baby' Holt, having a base of 1.63 m. (5 ft. 4 in.).
(2) The 'Baby Creeping- Grip,' having a base of 1.22 m. (4 ft.).

Of these two tractors the 'Baby' Holt was in use in the English Army as well as in the French, but it was not possible to obtain permission to use one as a basis of experiment for the land-ships. Only the 'Baby' Creeping-Grip remained, and this could not be obtained at once, as the tractor was still undergoing trials and experiment at the works at Chicago. For this reason, foreseeing the necessity of a longer base than that of the 'Baby,' Colonel Crompton ordered caterpillar tracks of the 'Baby' type lengthened and having a base of 2.75 m. (9 ft.) in May, 1915.

At first it was feared that the difficulties would recur that had already been found at the English manoeuvres in 1908, when the 'Caterpillar,' pivoting on one track, dug itself in till it grounded. It was evident that, given the weight of the armour, the armament and the crew, the total would amount to severa, [sic: several] tons which, even under the most favourable conditions [sic: conditions] would double the total load to be carried on the chain-track. The lengthened chain-tracks of the 'Baby' type were intended to serve for determining the proportions to be given to the chain-tracks for the actual land-ships.

[1] The secret report of Major-General E. D. Swinton was worded thus:- "Caterpillar machine gun destroyers. Suggested conditions to be adhered to in design, if possible. These are tentative and subject to modification.
Speed: Top speed on flat not less than 4 miles per hour; bottom speed for climbing (blank) miles per hour.
Steering: To be capable of being turned through 90 deg. on top speed the flat on a radius of twice the length of the machine.
Reversing: To travel backwards or forwards (equally fast).
Climbing: To be capable of crossing backwards or forwards on earth parapet 5 ft. thick and 5 ft. high, having an exterior slope of 1/1 and interior slope vertical.
Bridging: All gaps up to 5 ft. width to be bridged directly without dipping into them. All gaps above 5 ft. in width to be climbed (up to a depth of 5 ft. with vertical earth sides).
Radius of action: to carry petrol and water for 20 miles.
Capacity, crew and armament: To carry ten men, two machine guns, one light quick-firing gun."

The original idea of Mr. Churchill was that of transporting seventy infantrymen; the designs were commenced on February 25th, 1915, and were so advanced that an order for twelve land-ships was placed on March 20th, 1915, the land-ships to be 12 m. long (39 ft., the weight being 26 tons, and the engines 140 horse-power carried on pedrails on turn-table, and having a four-speed drive – see Fig. 1. At the same time the Admiralty gave an order to Mr.- now Sir William – Tritton for the construction of six double vehicles with large wheels 4.55 m. (15 ft.) in diameter, the drawings for which had been commenced by him in March, 1915, according to the original suggestion. Sir William Tritton maintained that, from actual experiments he had made, the slightest piece of barbed wire or other iron that became caught in the caterpillar tracks would cause jamming, whence the necessity for crushing barbed wire entanglements by means of large wheels. Before proceeding further, particulars were obtained as to the condition of the road in France behind the front, the radius of the curved, the strength of the bridges over the canals and rivers; and it was found necessary that the vehicle should be divided into two parts.

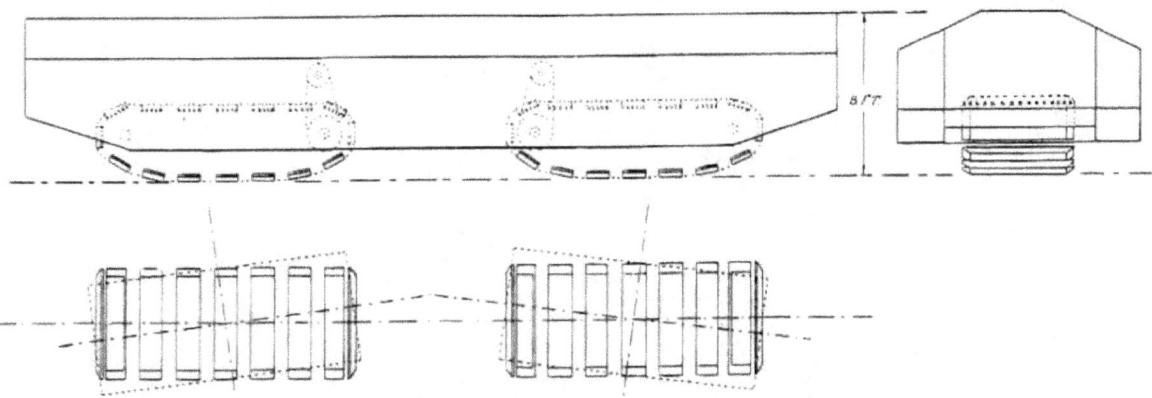

FIG. 1 – British landship to carry 70 men. (Turning circle 80 ft. 26 tons – 140 H.P.)
[additional information on the vehicles in Fig. 1 to 6 can be found in Hills, A. (2019). Pioneers of Armour 2: Colonel R. E. B. Crompton. FWD Publishing, USA]

The English land-ship divided into two parts now carried fifty-six men – Fig.2; the length of each of the parts of the vehicle was 6.70 m. (22 ft.), and the whole could turn in a radius of 12 m. (40 ft.) instead of the 24 m. (80 ft.) required by the first design.

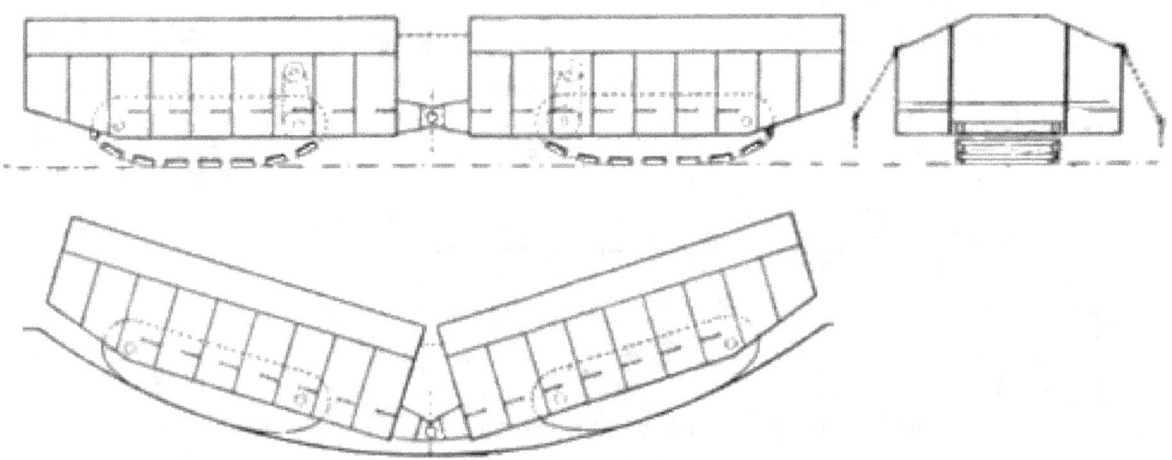

FIG. 2 – British landship to carry 56 men
(Pedrail tracks. Crossing shell hole 42 ft. diameter. Each half – 14 tons – 70 H.P.)

On account of the difficulty of obtaining the Pedrail tracks, other designs were commenced on May 8th for a land-ship similar to the second, but with 'Creeping-Grip' tracks – see Fig. 3. As the result of trials made with the Killen-Strait tractor in June, 1915 – crossing obstacles and barbed wire – the consulting engineers received the order to abandon designs for carrying infantry and proceed with offensive cars. In consequence of the difficulty in crossing bridges and turning curves, designs were commenced in July 1st for a double land-ship with two superposed turrets for 75 mm. artillery; the total height amounts to 2.88 m (9 ft. 6 in.) – see Fig. 4. On July 20th the height of this vehicle was required to be reduced, and this was done by abolishing the upper turret and reducing the total height to 2.29 m. (7 ft. 6 in.) – see Fig. 5. The height was still further reduced early in august to 6 ft. – see Fig. 6.

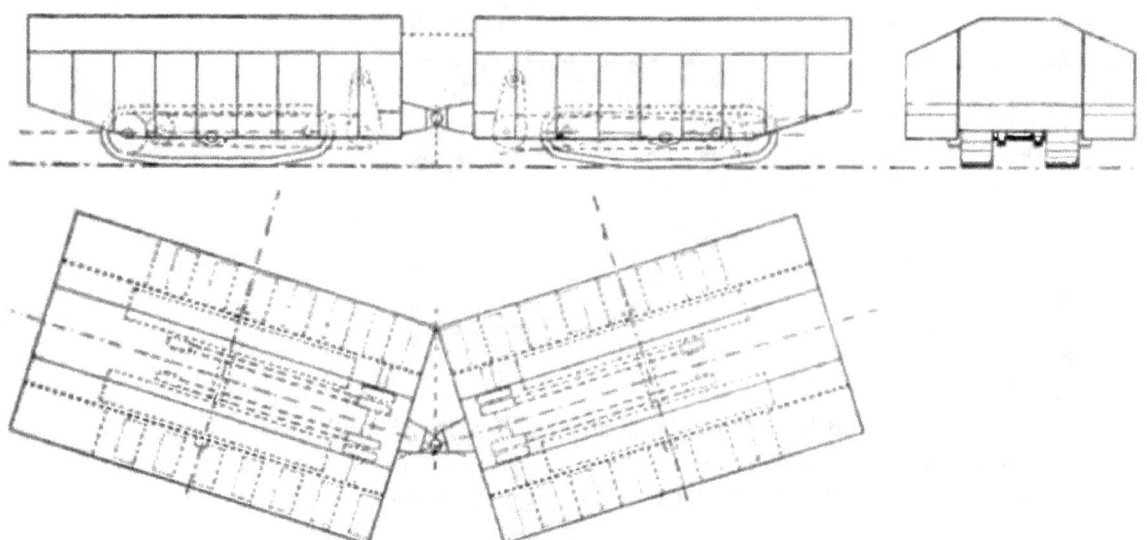

Fig. 3 – British landship with 'Creeping-Grip' tracks (Each half 14 tons – 70 H.P.)

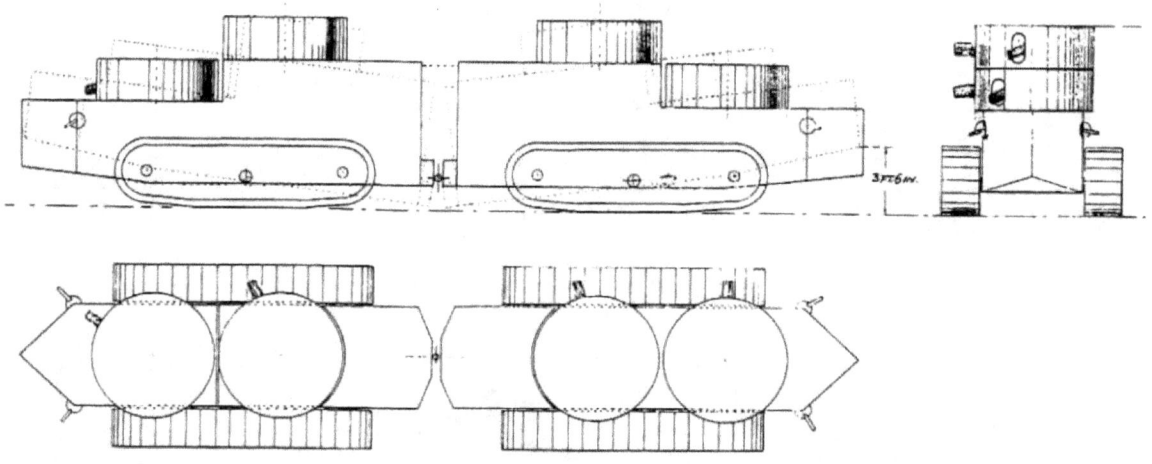

Fig. 4. – British double-turret gun-carrier, 9-feet 6-inches high.
(3-feet 6-inches elevation, Creeping-grip tracks. Each half – 14 tons – 70 H.P.)

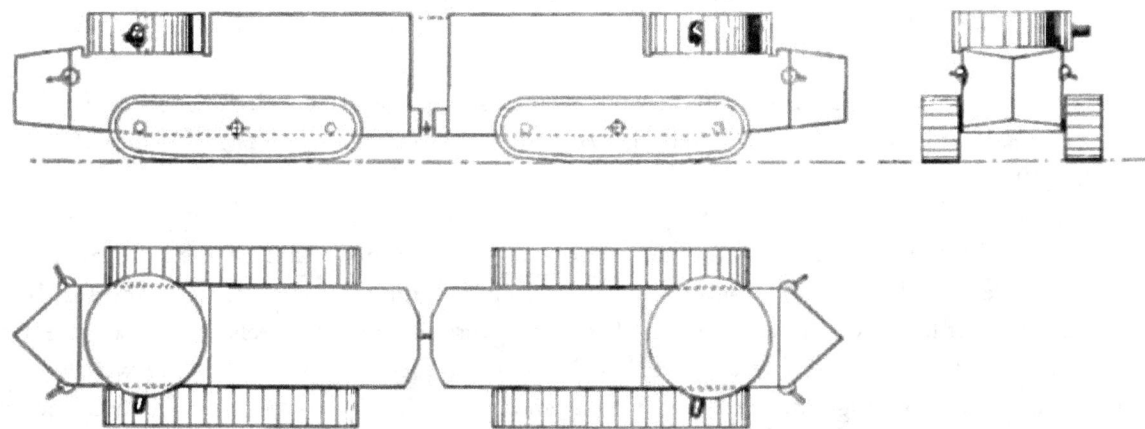

Fig. 5 – British single-turret gun-carrier, 7-feet 6-inches high. (Each half 13 Tons – 70 H.P.)

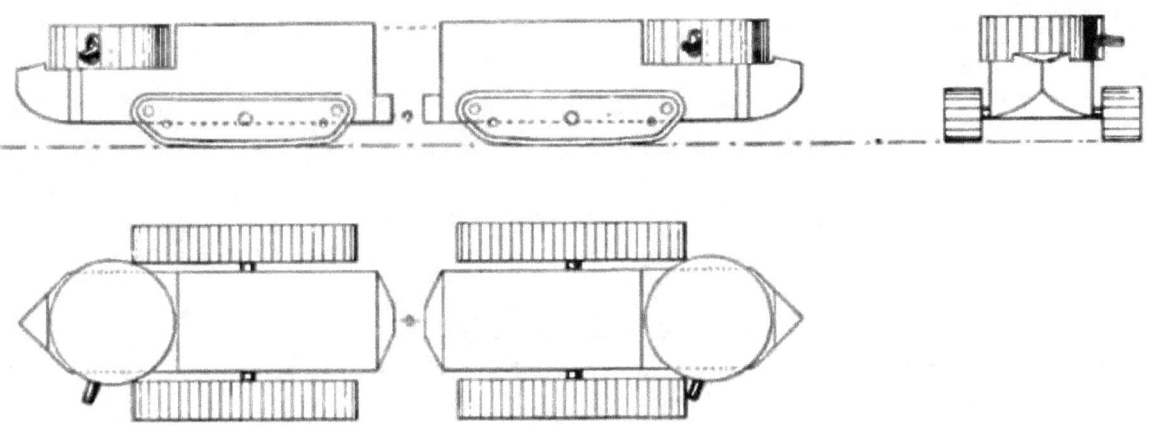

Fig. 6 – British single-turret gun-carrier, 6-feet high.
(Killen-Strait tracks. Each half – 12.5 tons – 70 H.P.)
[additional information on the vehicles in Fig. 1 to 6 can be found in Hills, A. (2019). Pioneers of Armour 2: Colonel R. E. B. Crompton. FWD Publishing, USA]

As the basis for calculations Colonel Crompton took the maximum pressure exerted by the chain-track at 550 gr. per square centimetre (8 lb. per square inch), the ratio of power to weight at a maximum of 5.5 horse-power per ton and a maximum speed of 7 kiloms. (4.4 miles) per hour; the results actually experienced showed that these figures were not only justified, but were exceeded before the end of the war. The trials made at Wormwood Scrubs before Mr. Lloyd George with the Strait tractor, and at Burton-on-Trent in August, 1915, with Creeping-Grip tractors gave inspiration to many inventors, among whom one may mention Messrs. Macfie, Wilson, Tritton and Nesfield. From these ideas were subsequently evolved those of the climbing and enveloping chain-track which are characteristic of the English tank. Sir William Tritton at that time proposed, moreover, to separate the power unit from the attacking unit, connecting them electrically by an armoured insulated cable. The idea was not proceeded with, but was later adopted in France by the firm of Saint-Chamond for heavy artillery.

THE BRITISH TANKS.[2]

[2] See account of the British tanks used in the war, by Sir Euston [sic: Eustace] H. Tennyson d'Eyncourt, K.C.B., read before the British Association, 1919, and published in *Engineering*, September 12th and 19th, 1919.

The chain-track offensive machines were called during the period of the construction in England 'tanks,' to distract attention from their object; this rather inappropriate name has been retained. The first tanks were put in hand in 1916. They were fitted with wheels at the rear, intended for facilitating steering and balance – see Fig. 7; later experience showed that the wheels were unnecessary. The first design included the climbing and enveloping chain-tracks passing over the top of the vehicle. It also included two lateral overhanging turrets (sponson) removed for transport. Various arrangement for change of speed were adopted; in the first instance the sliding gear change was used, but later the epicyclic gear was adopted in the marks built in 1917 and 1918; trials were also made with hydraulic and electric transmission.

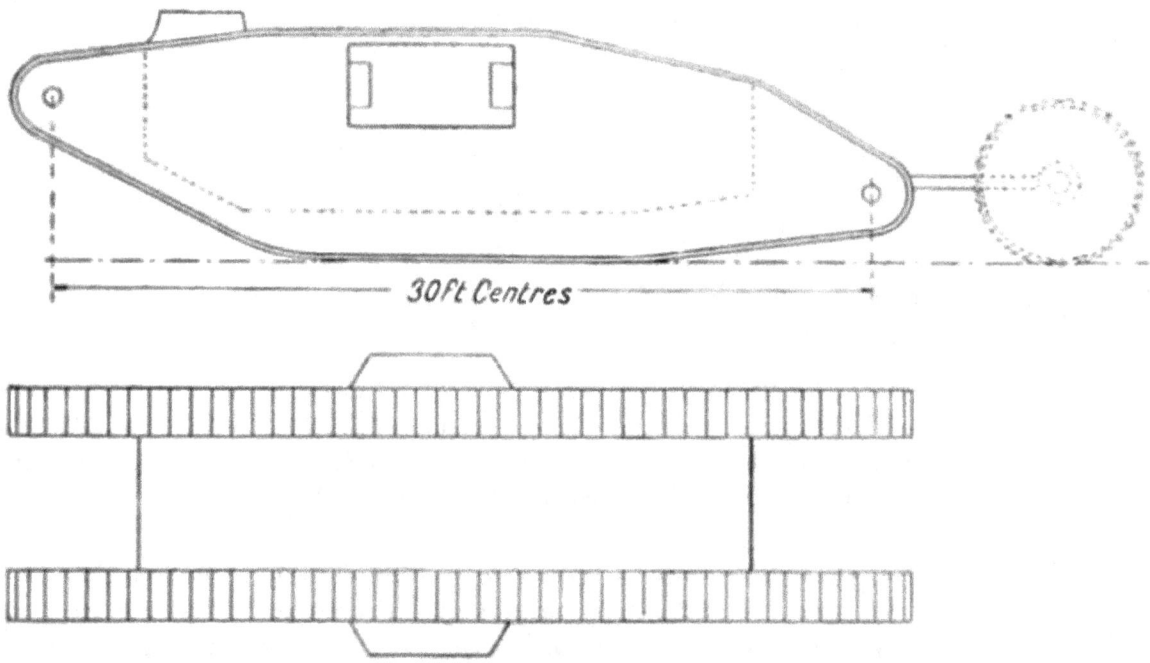

Fig. 7 – Original British tank with tail wheel. (26 tons 105 H.P. 1916)

The 'Whippet' (Medium Mark A) – Fig. 8 – a high-speed machine-gun tank weighing 14 tons, was fitted with engines of 90 horse-power. Its maximum speed was 14 kiloms. (8.7 mils) per hour. In this mark, the chain-track is not of the enveloping pattern used in several British marks. Later an attempt was made to combine the principle features of the two first types in Mark C – Fig. 9. Finally, the International type (Mark VIII.) – Figs, 10, 13 and 14 – shows the final type of 1918.

The first tanks revealed some defects due to want of appreciation of the essential conditions to be fulfilled:-

(a) Ability to run over an hard projecting obstacle; involving the concentration of the whole of the load on one point of the chain-track.
(b) The angle-guides for the chain-track shoes; these had been arranged to hold up the lower portion of the chain-track in passing over trenches; it was found later that this arrangement was not only useless, but the cause of breakdown and of difficulty in dismantling the chain.
(c) Absence of springing, limiting the speed.
(d) In the early marks the co-operation of three or four persons was required for manoeuvering; this made the driving extremely difficult.

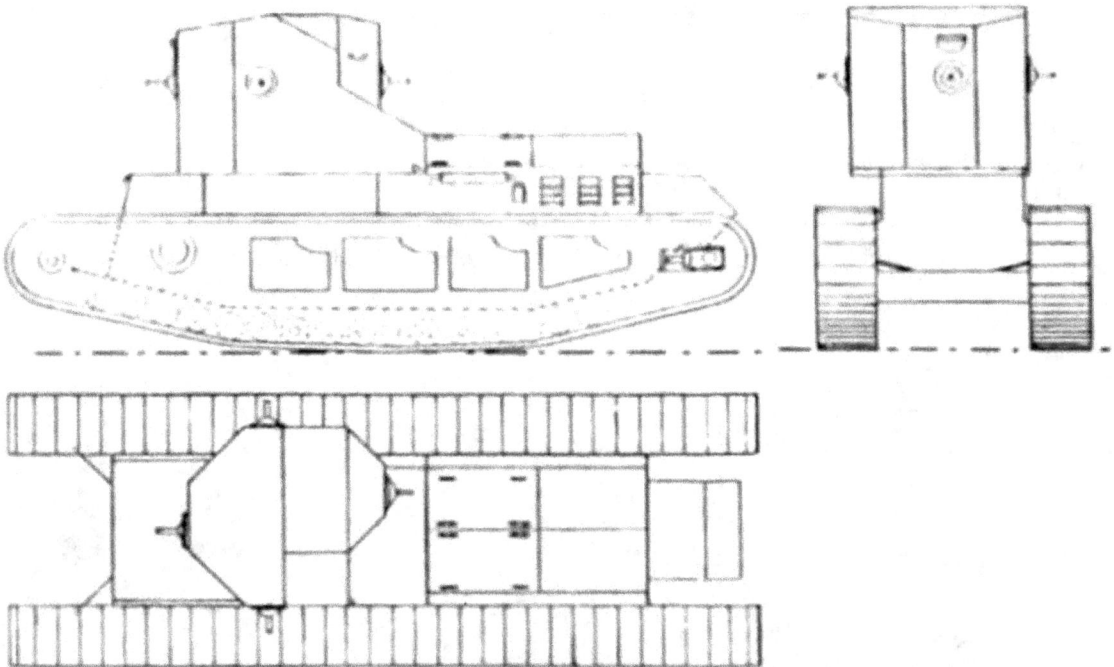

Fig. 8 – British 'Whippet' tank – Medium Mark A. (13 tons – 48-80 H.P. 1918)

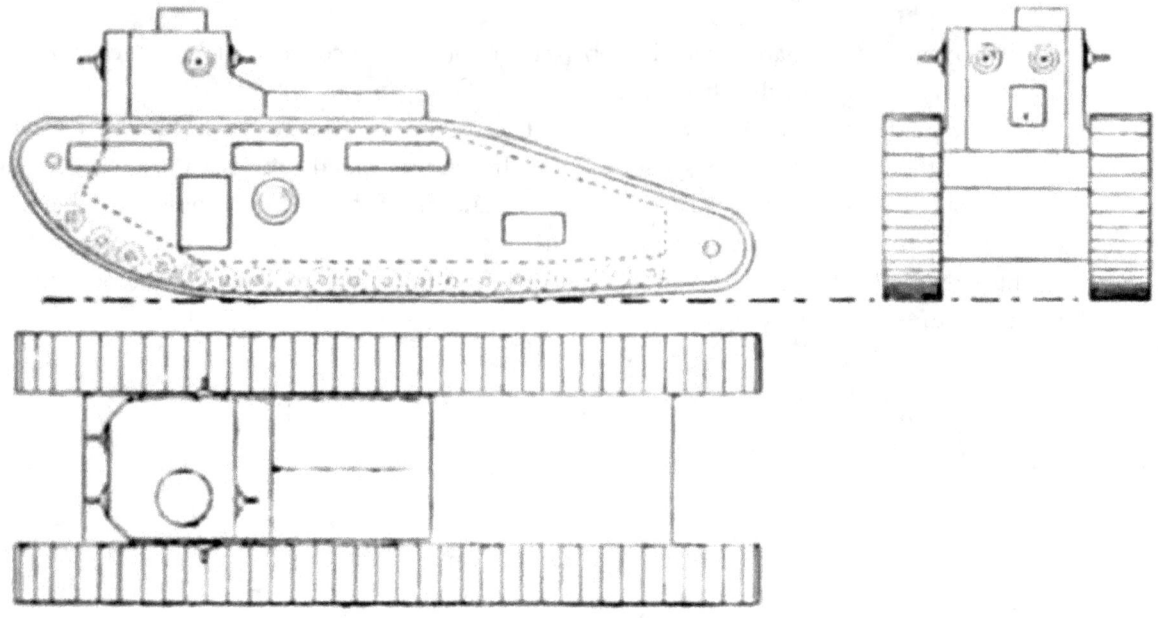

Fig. 9 – British tank – Mark C. (18 tons – 150 H.P. 1918)

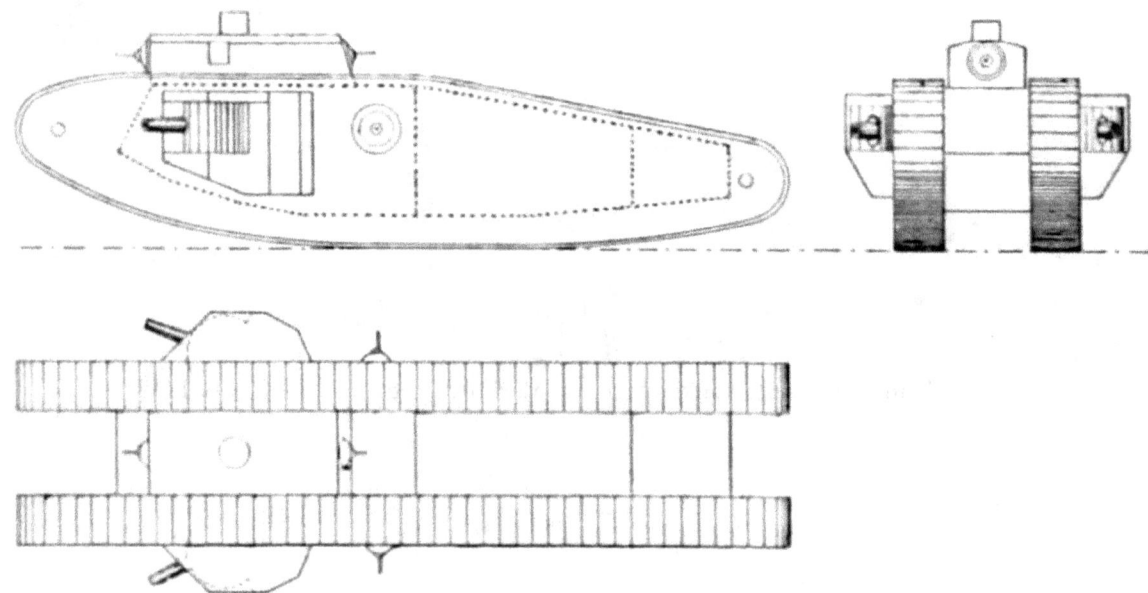

Fig. 10 – International tank – Mark VIII. (30 tons - 300 H.P. 1918)

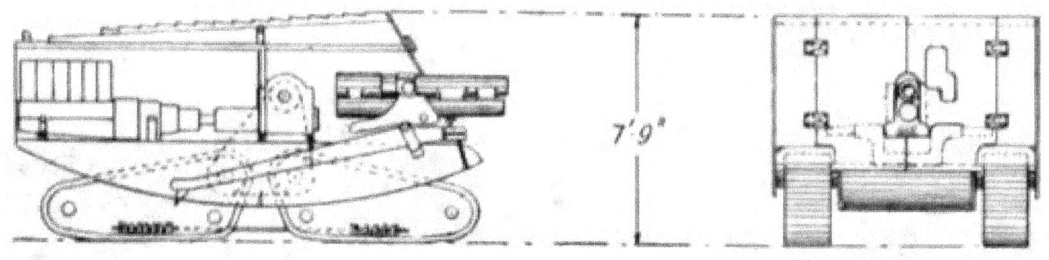
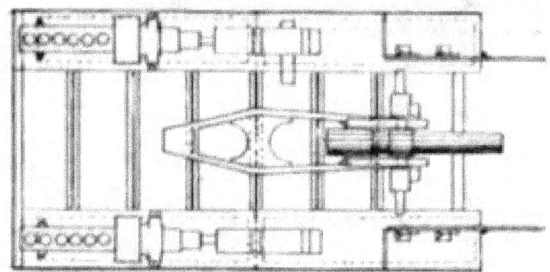

Fig. 11 – British armoured motor gun-carrier. (12 tons – 110 H.P.)
[additional information on this particular vehicle can be found in Hills, A. (2019). Pioneers of Armour 2: Colonel R. E. B. Crompton. FWD Publishing, USA]

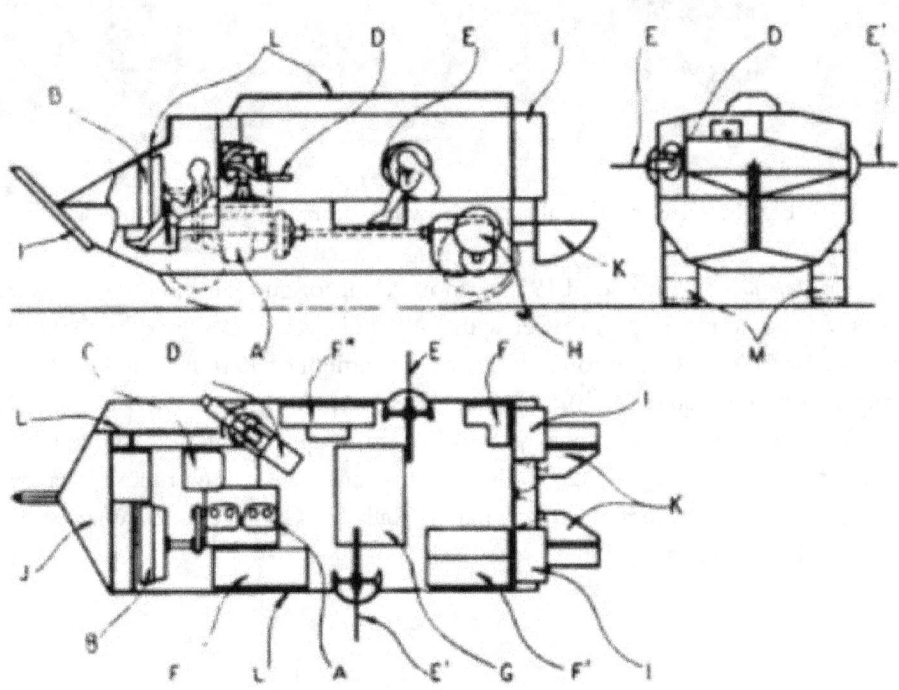

Fig. 12 – Schneider 'char d'Assaut'.

A – Motor, B – Radiator, C – Drivers [sic: driver's] cabin seat, D – Short 75 mm short gun, E. E^I – machine guns, F. F^I. F^{II} – munitions, G – Well in floor, H – Gearing, I – Petrol tanks, J – Nose, K – Rear wings, L – Armour, M – Chain Tracks.

Fig. 13 – International tank.

Fig. 14 – International tank, showing internal arrangements.

Other War Material.- At the end of 1915, armoured motor gun-carrying tractors were suggested. Fig. 15 shows an armoured motor gun-carrier with a 4.5 in. howitzer designed by Colonel Crompton. In the design made by Colonel Crompton at the commencement of 1916 the total load was 600 gr. per square centimetre (9 lb. per square inch), the ratio of horse-power to weight 9 horse-power per ton, and the speed 13 kiloms. (8 miles) per hour, with two motors capable of being coupled and disengaged and each fitted with epicyclic change gear. Gun-carrying tanks were constructed later by Major J. R. Greg, C.B.E., of the Metropolitan Amalgamated Railway Carriage and Wagon Company, in collaboration with Major Wilson.

THE FRENCH TANKS.

SCHNEIDER.

As has been stated above, the first French tanks were built by the firm of Schneider. The 400 (C.A. Type) tanks that has been ordered from the on February 25th, 1916, were of 13 tons weight. The first batch was delivered in September of the same year. See Fig. 15 on page 168 of our issue of December 2nd, and Fig. 15. The armament of these cars consisted of a 75 mm. short gun in front and two machine guns at the sides. The motor was of 60 horse-power, four cylinders 135 mm. by 170 mm. (5.3 in. by 6.7 in.), three-speed change gear with sliding gears, speed s from 2 kiloms. to 8 kiloms. per hour (1.2 to 5 miles per hour); steering was effected by declutching and braking on of the chain tracks. The arrangement of the chain-tracks comprised two jointed truck frames on each side carrying the suspension springs with a transverse compensating lever in front. The crew consisted of seven men. Apart from the tanks, the firm of Schneider constructed other chain track artillery during the war, as follows:-

(a) *Tractor Lorries*, of the C D type, capable of carrying a load of 4 tons, and giving a draw-bar pull of 7 to 8 tons. These tractors used for the greater part the mechanical details of the C.A. tank, except that the change-speed gear had four speeds. The motor was the same, as also was the arrangement of the chain tracks.

(b) *Carrier Tractor,* of the C D 3 type, capable of carrying a 155 mm. gun either long or short, or a 220 mm. howitzer. This vehicle was fitted with a crane and a capstan, as well as the gear necessary for enabling it to pick up its load in a few minutes. Weight 12 tons. Motor the same as in the preceding type C D; speeds from 1.2 kiloms. to 6 kiloms. (0.6 to 3.7 miles) per hour; draw-bar pull in first speed, 10 tons.

(c) *Self-propelled Chain Track Mounting for 220 mm Long Gun.* – The gun, which can be fired at a maximum angle of 37 deg. to the horizontal – see Fig. 16 – is carried on its small normal type of mounting, which rests on two inclined roller paths forming part of the chassis of the vehicle. A hydraulic brake is interposed between the small mounting and the chassis. The return to loading position is effected by gravity. The motor was six-cylinder of 120 horse-power; total weight, 40 tons. The gun was laid roughly for direction by means of the chain tracks, traversed by the main motor, and the fine adjustment effected either by a hand wheel or by a small auxiliary 10 horse-power motor driving through a high-ratio reduction gear.

SAINT-CHAMOND.

In 1915 also the Forges et Aciéries de la Marine et d'Homécourt, under the direction of Colonel Rimailho began experiments and designs for a tank, the first example of which was commenced in May, 1916. Trials of this took place in July, and the manufacture of 400 tanks of this pattern – see Fig. 17 – was commenced in November of the same year. This tank was fitted with a motor of 90 horse-power, running at 1450 revolutions per minute, of the four-cylinder, sleeve-valve Panhard type. The total weight of this tank in running order, without munitions or crew, was 19.9 tons, and with munitions and crew 21.5 tons. One of the most important characteristics of this vehicle consisted in the electric transmission with infinitely variable change of speed which avoided the great difficulty in changing gear of the early English tanks.

The armouring was as follows:- On the sides two plates having a total thickness of from 15 mm. to 17 mm.; in front, 11 mm. plate inclined at 45 deg.; at the back, 8.5 mm plate; and for the roof, 5.5 mm. plate. The total length of the tank, exclusive of the projecting barrel of the gun, was 7.91 m. (26 ft.), and its width was 2.67 m (8 ft. 9 in.); the maximum height was 2.365 m. (7 ft. 9 in.); the ground clearance, 0.500 m. (19.7 in.); the width of the chain track was 0.500 m. (19.7 in.); the length of bearing with 0.050 m. (2 in.) sinkage was 3.350 m. (11 ft.).

This tank carried a 75 mm. quick-firing Saint-Chamond gun or an 1897 pattern 75 mm. gun, with its supply of cartridges. The contents of the petrol tanks were 265 litres (58 gallons), distributed among three petrol tanks. This supply was sufficient for running a distance of 35 kiloms. (22 miles) on slow speed and of 60 kilom. (37 miles) on top speed. Steering was effected by means of a steering hand wheel acting on contactors, the closing or opening of which enabled different combinations of the internal connections of the electric motors to be made or broken. These combinations acted by accelerating the outer motor and slowing, braking or locking the inner motor on the curve.

Fig. 15 – Schneider 'Char d'assaut'

Fig. 16 – Schneider long 220 mm. gun-carrier

It is of interest to note that the Saint-Chamond firm had the idea of using a carrying chain track in the front of the vehicle – see Fig. 18 – for assisting in climbing obstacles, as had been proposed by Colonel Swinton. This design, however, was not actually executed. The Saint-Chamond firm also designed and built carrying tractors for the 120 mm. long quick-firing gun, as well as a motor chain track gun mounting carrying a 120 mm. long type long-range gun. The large chain-track artillery constructed by the Saint-Chamond firm consisted of two chain-track vehicles per unit, and is particularly worth attention. The two vehicles in running position are in both cases carrying a 194 mm. Mark F gun – shown in Fig. 19, and the gun in firing position is shown in Fig. 20. In another instance a 280 mm. short howitzer was carried – see Fig. 21.

Fig. 17 – St. Chamond tank

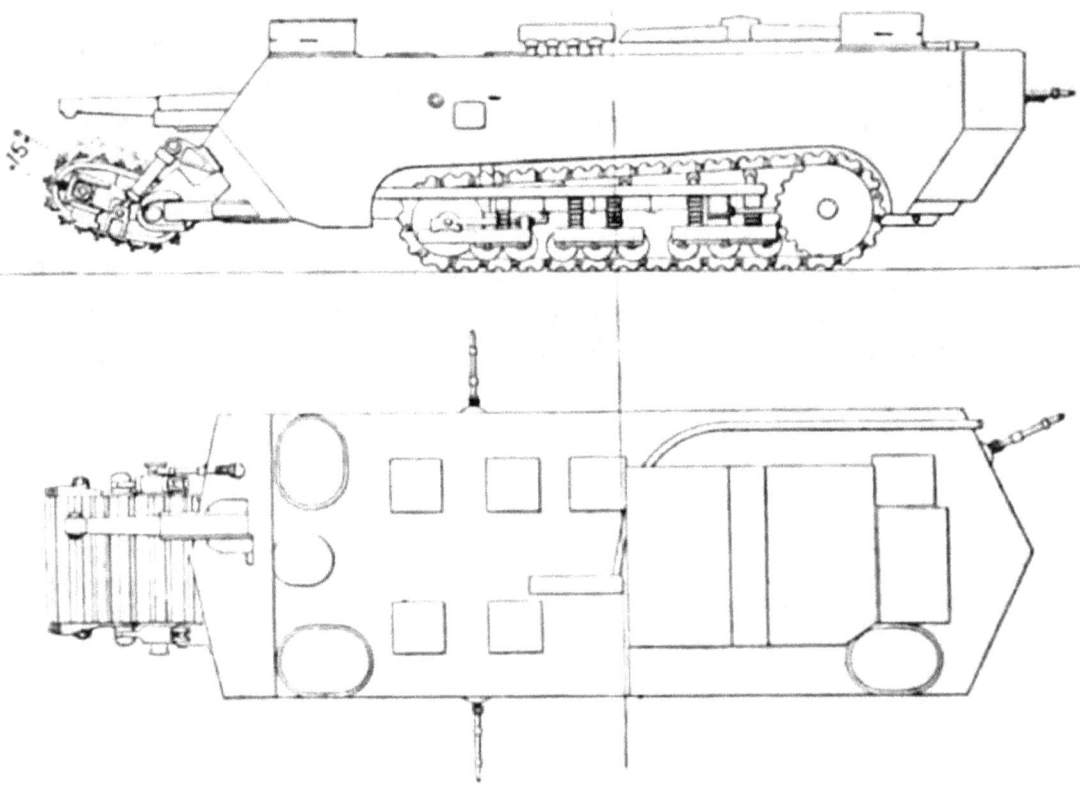

Fig. 18 – St. Chamond tank, with subsidiary track in front

Fig. 19 – St. Chamond petrol-electric car hauling gun

Fig. 20 – St. Chamond electric car with gun in firing position

Fig. 21 – St. Chamond electric car with 280 mm. howitzer

The unit in each case consisted of two chain-track vehicles. The forward motor chain-track vehicle, fitted with an electric generating set carried the ammunition. The chain-track gun mounting was fitted with two electric motors, and supplied from the generating plant on the forward vehicle by means of a cable 50 m. (55 yards) long. On uneven country the vehicles were generally run independently of each other. If the slope or condition of the ground required it, the two vehicles could be moved one after the other, and thus make progress by successive stages.

RENAULT.

Following the particulars given by General Estienne, the firm of Renault in 1916 began the designs of a light tank capable of being carried on a lorry. The first trials of this were made in March, 1917, and mass production followed immediately. The tank is shown in Fig. 22 and 23. The Renault tank was fitted with climbing tracks, and had an 18 horse-power four-cylinder motor with the change and driving gear at the rear of the tank. The armament consisted of a machine gun carried in a revolving turret. The suspension gear was arranged at the ends of the chain track frames, which were independently sprung. The vertical armour consisted of 16 mm. plate; plates slightly inclined to the vertical were 8 mm. in thickness; the plates, horizontal or slightly inclined to the horizonal, were 6 mm. in thickness. The method of springing the chain-track frames consisted of plate springs, compensating levers and bogies, which ensured a sensibly constant and uniformly distributed load over each of the carrying rollers, whatever might be the irregularity of the ground to which the chain must adapt itself. The tank was controlled entirely by one man. The turret, carried on a ball race, could be swung round its vertical axis, and the machine gun or quick-firing gun could cover the whole of the horizon. Steering was performed by unclutching the chain drive to the side to which it was desired to turn, and by locking it if required to turn on the spot. To facilitate the running of the tank over trenches, a removeable tail was fitted behind the armouring. The average weight was 6.7 tons.

THE AMERICAN TANKS.

When America entered the war very little authentic information covering the use and types of tanks was available in that country. Some rather vague specifications were obtained from France during the summer of 1917, from which two patterns of experimental tanks were built. One of these was steam driven, the other equipped with a petrol-electric drive. Before the experiments with these had been concluded, the great importance of the tanks had made itself apparent. The 'Ordnance Department' of the United States then sent one of its officers to Europe on a special mission with the object of securing all available information relating to the construction and use of the tank. After numerous conferences with the British and French authorities, America undertook the construction of two types of tanks for the American Army – the small two-man tank, based on the French Renault tanks, and the large 30-ton tank to be produced jointly with England.

Renault tanks were purchased and sent to America, with a complete set of drawings, and instructions were given to duplicate these tanks, using American standards of manufacture and measurements. The designs and specifications relating to the large Anglo-American Mark VIII. Tank – see Figs. 10, 11 and 12 *ante* – were completed after numerous conferences with the British General

Staff, and the orders were then put in hand for simultaneous manufacture in England and America. According to a treaty signed by the two countries on January 22nd, 1918, the commissioners appointed under this treaty were empowered to complete the designs, to arrange for the production of British and American components, build a factory in France for the assembly of these components, and actually assemble the tanks. The designs were commenced in November, 1917, and by May, 1918, practically all of the drawings had been received in America, where the production of parts, including motor, transmission, radiators and chain-track rollers and sprockets was to take place. The British Government undertook to supply the chain track, armour, framing and armament. The British works had greater experience in the production of armour plate, and possessed a large stock of guns and other fittings necessary for the armament of the tanks.

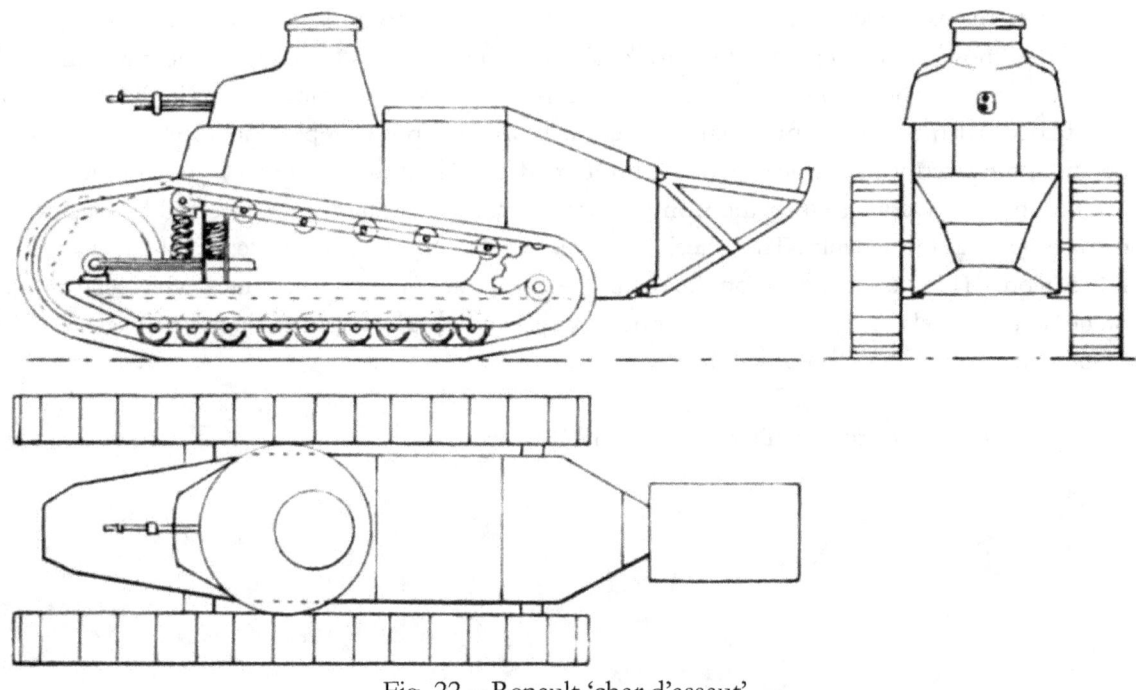

Fig. 22 – Renault 'char d'assaut'
(6.5 tons 18-39 H.P. 1918)

During this period the construction of American tanks of the Renault type – see Fig. 24 – proceeded, and the assembly was entrusted to three large American works – the Van Dorn Ironworks, Cleveland. Ohio; the Maxwell Motor Company, Dayton, Ohio; and the C. L. Best Company also of Dayton. The order was for 4400 tanks, of which only 950 were actually completed. During the summer of 1918 work was continued on the Mark VIII. tanks. A sample set of British components was sent to America and, with the addition of the American components which had been quickly prepared, the assembly was carried out of a complete tank, which was tested at the works of the Locomobile Company, Bridgeport, Conn. Everything was in good order, both in England and in United States, as well as at the American works at Neuvy-Pailloux, in France, which was ready for full production when the Armistice was signed. The manufacture of 1500 tanks was then in hand, with a view to a heavy delivery in the spring of 1919.

America, moreover, had undertaken the construction of 1450 complete tanks from parts manufactured entirely in the United States, in addition to the components which were already being made in the United states. During the summer and autumn of 1918, a 3-ton tank – see Fig. 25 – much smaller than the Renault, was designed by the 'Ordnance Department,' to be constructed by the Ford Motor Company, of Detroit. These designs made use as far as possible of the standard parts of the Ford automobile. Just before the signing of the Armistice, orders were placed for 15,000 of the 3-ton tanks, which were intended to be used either as light tractors or as machine gun tanks. America had completed 100 Mark VIII. tanks in the spring of 1920 for the service of the 'American Tank Corps.'

Chain-track artillery has been studied in America‡ on principles identical with those adopted in France. Experiments made by the firm of Holt as far back as 1916 showed the possibility not only of carrying a 200 mm. howitzer but also of firing it from the chain-track mounting. More than six definite patterns have been constructed, the armament of which comprises the 155 mm. long gun and the 240 mm. howitzer carried on automotor chain-track mountings. Of these, the most remarkable is the Christie automotor mounting, having four wheels at each side, 915 mm. (36 in.) in diameter, and a chain track carried below the mudguard and capable of being put into place in 15 min. For this last operation, the intermediate wheels which on the road are raised out of contact with the surface are lowered, and the steering gear of the front wheels is locked. Steering is then effected by declutching on chain track, as in the tanks. The wheels are fitted with solid rubber tires [sic: tyres], giving a speed of 27.5 kiloms. (17 miles) per hour on the road. The chain track is fitted with snugs, which engage with notches in the wheels between the solid rubber tires [sic: tyres]. The speed attained on the chain tracks is about 14.5 kiloms. (9 miles) per hour.

‡ *Journal* of the United States Artillery, January, 1921.

Fig. 23 – Renault 'char d'assaut'

Fig. 24 – American tank of the Renault type

Fig. 25 – 3-ton tank made by the Ford Company

THE ITALIAN TANKS.

By the order and at the expense of the Italian Government, designs for the construction of tanks were undertaken by the Fiat firm at Turin – see Fig. 26. The tank was of heavy pattern, weighing 35 tons, with a 250 horse-power motor and a maximum speed of 12 kiloms. (7.4 miles) per hour. The armament consisted of a short gun carried in a revolving turret.

The Fiat firm is at present engaged on the construction of a lighter tank – see Fig. 27 – the weight of which is about 6 tons. The motor is of 45 horse-power, running at a normal speed of 1500 revolutions per minute, and giving a maximum speed of 16 kiloms. (10 miles) per hour. The armament consists of two coupled machine guns arranged in a revolving turret.

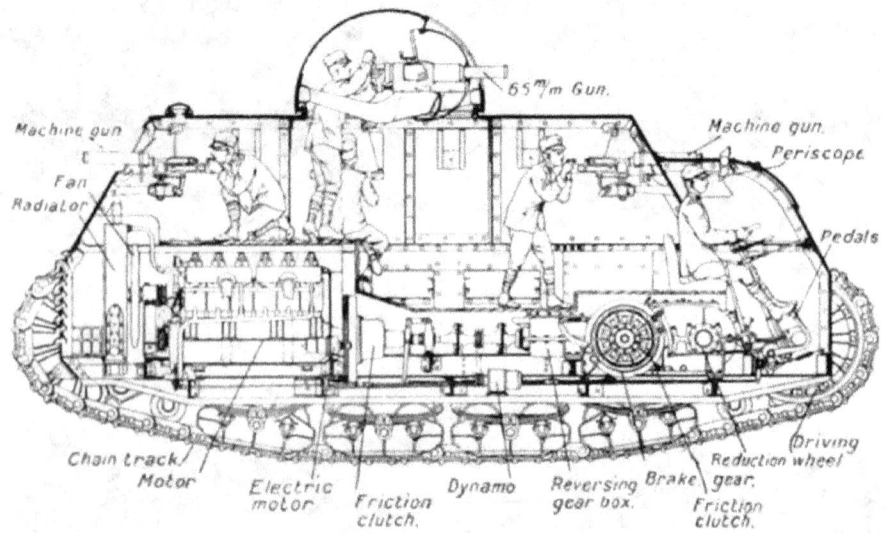

Fig. 26 – Interior of the heavy pattern Fiat tank

Fig. 27 – Interior of light pattern Fiat tank

THE GERMAN TANKS.

It was not till near the end of the war that Germany began the construction of tanks. The type produced by them bears no trade of the improvements which had been gradually introduced into the design of this engine of war by the Allies. It is more particularly in methods of defence against tanks that the Germans made the greatest progress in this branch of warfare:-

(1) By the adoption of a gun taking a cartridge almost double the linear dimensions of the ordinary infantry rifle cartridge.

(2) By the construction of hidden excavations modelled on the elephant pit, with the object of trapping the tanks which might run over the prepared ground.

THE ESSENTIAL DATA FOR DESIGNING CHAIN-TRACK VEHICLES.*

(1) *Gearing with the Ground (Tangential Effort)*.- The link of the chain track is usually formed with two projections which act as teeth and gear with the slabs of earth between them. Resistance to traction can therefore be measured in proportion to the area of these slabs of compressed earth imprisoned between the consecutive projections on the chain track. As soon as the chain track begins to rise the compression of the earth and the resistance to shearing diminish; the area of the front and central part of the chain track in contact with the ground can therefore only be taken into account in determining the effective resistance.

On grass land the tangential resistance under normal conditions of loading may be taken as 3 kilos. per square centimetre (45 lb. per square inch), but on clayey soil, as soon as the shearing commences the slabs fill the spaces formed between the projections on the chain links, thus producing a uniform smooth surface which can slide indefinitely on the ground. For crossing such ground it is necessary to use detachable spuds (called grousers in America), which by increasing the depth of hold on the ground, enable the necessary resistance to be obtained. According to the results of experiments made with agricultural tractors it appears that spuds of good shape may transmit a load of 3 to 5 kilos. per square centimetre of the vertical area (45 lb. to 75 lb. per square inch). Marshy ground so soft as to require a reduction in the insistent load is dealt with by an increase in the area of contact of the chain track; this is obtained by widening the chain track and consequently enlarging the slabs of earth which resist shearing. As steep gradients are not to be feared on such ground, the use of spuds is unnecessary.

(2) *Insistent Weight, Load per Square Centimetre (or per Square Inch)*.- It is essential that a tank should be able to travel over ground over which not only cavalry can pass, which requires a vertical resistance of the ground amounting to 2 kilos. per square centimetre (30 lb. per square inch), but also infantry, which only requires a vertical resistance of about 500 to 600 grammes per square centimetre (7.5 lb. to 9 lb. per square inch). It is even required that they shall go where infantry cannot pass; it is mentioned that some American chain-track tractors have been modified in certain instances by widening the chain track so as to be able to cross marshes capable of carrying a load of only 150 grammes per square centimetre (2.25 lb. per square inch).

The difficulties of traction are still further increased when snow is in question. It is interesting to note that the ski runner requires a vertical resistance of only 50 grammes per square centimetre (1.5 lb. per square inch).

The carrying surface increases considerably with the depth of sinkage, but not in the same proportion with the caterpillar as with the wheel. An easy method exists for increasing the bearing surface by about 30 per cent., if the nature of the ground so requires, by the use of chain links widened on the outer side so as to form a cantilever projection. Non-symmetrical chain tracks were used in several instances on the Saint-Chamond and English tanks.

(3) *The Power per Ton.-* The power per ton of commercial tractors varies from 8 to 15 horse-power. In the case of the tank the weight is increased by the armouring, and the running conditions are rendered still more difficult because the tank is required to climb gradients which may be as steep as 45 deg. In such a case the measurement of the gradient must be taken, for calculations relating to the tractive effort, as the sine of the angle instead of the tangent; the necessity for doing this does not occur on the steepest gradients found on ordinary roads, and still less on railways.

The maximum effort of traction of agricultural and commercial chain-track tractors generally amounts to from 35 to 60 per cent. of the weight. The figure reaches 80 to 100 per cent. in the tanks, permitting them to climb very steep gradients which in some cases have even exceeded 45 deg.

(4) *Limiting Dimensions.-* the width and height of the tank are limited by the requirements that on the one hand the vehicle must be able to run on roads, and on the other hand that it must be capable of being carried by rial on a flat-topped wagon. The height of the tank is not only limited by the loading gauge, but also by the necessity for keeping the visibility of the vehicle as low as possible. Only in a few instances have the tanks exceeded 2.50 m. (8 ft. 2 in.) in height.

(5) *Speed;-* The speed of the commercial tractor varies in the case of heavy tractors from 1.7 to 5.5 kiloms. (1 to 3.5 miles) per hour; in the case of fast tractors from 2.6 to 12 kiloms. (1.5 to 7.5 miles) per hour.

The first English tanks (Marks I. to IV.), which were very heavy, had a speed of only 1.2 to 6 kiloms. (0.75 to 3.7 miles_ per hour; the Schneider tanks had a speed of 2 to 6.7 kiloms. (1.2 to 4.2 miles) per hour; and the Renault tanks 1 to 7.8 kiloms. (0.6 to 4.8 miles) per hour. The fast English 'Whippet' tanks (medium Mark A) took more account of the principle formulated by Colonel Crompton that "real armouring is speed"‡, and attained a speed of 2.4 to 13.5 kiloms. (1.5 to 8.3 miles) per hour.

‡ This expression was also adopted by Lord Fisher, Chief Lord of the British Admiralty, in regard to the armouring of battleships.

(6) *The 'Effectiveness,' that is to say, the Immunity from being Struck by Enemy Fire.-* As a base on comparison between designs, Colonel Crompton laid down the principle that the "effectiveness of a tank is a function of the square of its speed divided by its height"; this principle has been largely justified by the changes of design in the actual tanks produced, but

it could be reduced to definite terms by experiments in firing at targets of the form and profile drawn across uneven ground.

(7) *Control and Steering by One Man.-* Attention has already been drawn above to the fact that the control and braking of the tank should be in the hands of one man only; this is essential for rapidity and safety in manouevering the ordinary automobile, and is still more necessary with the extremely sharp turnings that can be made with the tank.

(8) *Armouring.-* The thickness of the armour should vary, as on battleships, according to the portion of the tank to be protected and according to the angle which the surface presents to enemy fire. At the commencement it was found that a thickness of armour of 10 mm. would stop the German bullet, but it was soon found that at short range and with the reverse rifle bullet, as used by the Germans for this object, a thickness of 11 mm. and even 12 mm. (0.43 in. and 0.47 in.) was barely sufficient.

(9) *Position of the Centre of Gravity.-* The position of the centre of gravity is of great importance in determining the ability to climb obstacles. A high position of the centre of gravity requires a greater inclination of the tank before equilibrium is reached on the edge of the obstacle – Fig. 28. This effect may also be produced by a lower position of the centre of gravity, but more to the rear of the tank, which brings it on the same critical line; there is another reason for keeping the centre of gravity towards the front of the tank; in every automobile vehicle, when running forward, there is a couple exercised by the motor and increased in magnitude by the gear between the crank shaft and the periphery of the wheel gearing with the chain track, the effect of which is to produce a virtual [sic: virtual] displacement of the centre of gravity towards the rear – Fig.29. It will be remembered that in the first automobiles, when climbing hills, the front wheels sometimes lifted and that the vehicle could turn over about the back axle instead of progressing. It must not be forgotten that in descending gradients this couple changes its sign, and that a position of the centre of gravity too far forward would involve the lifting of the rear of the car and cause turning over forwards – Fig. 31.

(10) *The Curve of the Chain Track in Contact with the Ground.-* The first chain tracks made in France were based on the Holt system, with spring suspension.

The English tanks had no spring suspension; the first tank has a very flat and very long bearing surface on the ground, but the later marks took a more oval form, with very short, straight lengths and gradual changes of curvature. Although this form does not give the minimum sinkage, it has numerous advantages for control and steering.

(11) *The Path Traced by a Point on the Caterpillar.-* This question is of interest in considering the ability of the vehicle to climb projecting obstacles.

In the case of the Holt type of chain track – Fig. 30 - with equal semi-circular ends, any point follows a path which consists in two cycloids connected by the common tangent of a length equal to double that of the straight part of the chain in contact with the ground.

In the case of the Renault type of chain track having its ends formed as two circles of different diameters, the curve takes the form of two portions of different cycloids connected by a common tangent.

When the form of the chain track is arbitrary the curve must be considered graphically.

A study of these curved will enable one to take account of the variations in the propulsive effort against an obstacle and in the resistance offered by the ground.

In the case of a tank in contact with a vertical obstacle – Fig. 32 – the point B of the frame becomes displaced to B¹; the chain track remains in gear with the ground, while the point A at the front climbs to A¹.‡ The resultant of the shearing effort on the ground and of the weight of the tank tends to overturn the obstacle.

‡ This was originally written as "*the point A of the frame becomes displaced to A¹; the chain track remains in gear with the ground, while the point B at the front climbs to B¹*" when published on 16th December 1921 in The Engineer. This was corrected by Legros in a letter published 23rd December to "*the point B of the frame becomes displaced to B¹; the chain track remains in gear with the ground, while the point A at the front climbs to A¹*"

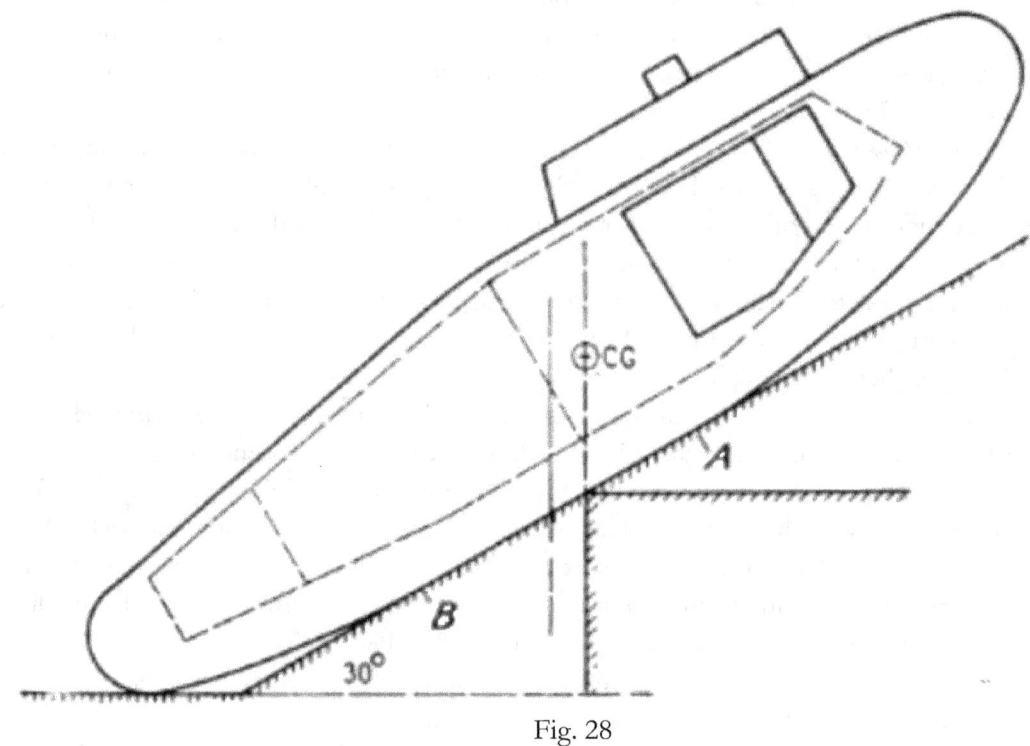

Fig. 28

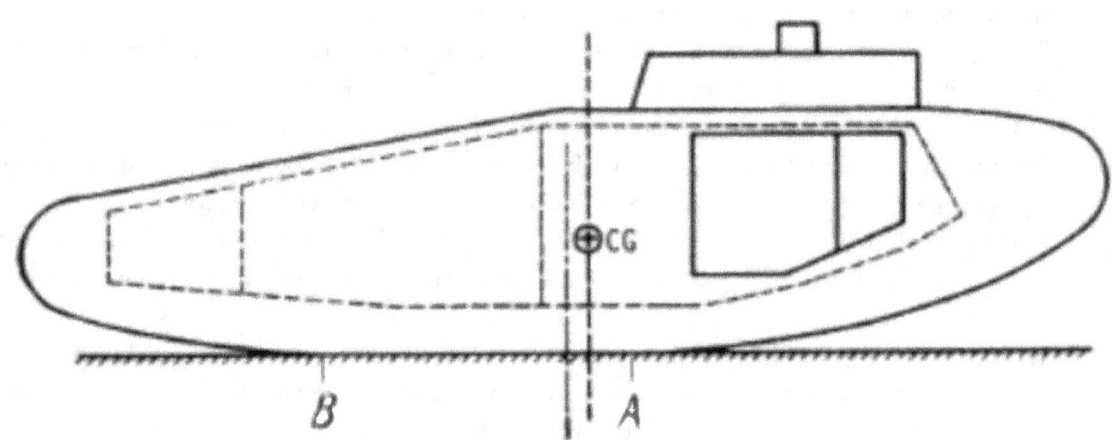

Fig. 29

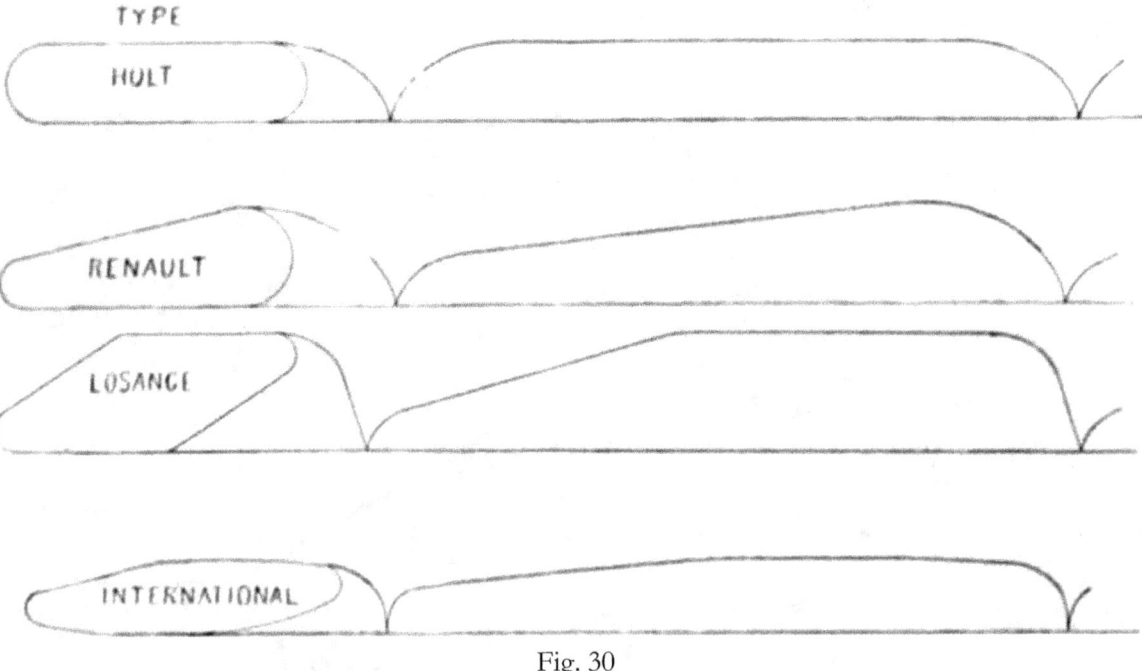

Fig. 30

(12) *Distance Run.-* The distances actually run by tanks on roads and on the ground are widely different; according to the observations made by General Estienne during the war, out of a total distance of 500 kiloms. run by a tank, only 30 kiloms. were run off the roads. Now, running on the road wears out the chain track much more than running over soft ground. Numerous trials and experiments have been made in France, in England, and in America to fit the vehicle either with wheels or with removeable chain tracks. Up to the present no system has shown distinct superiority.

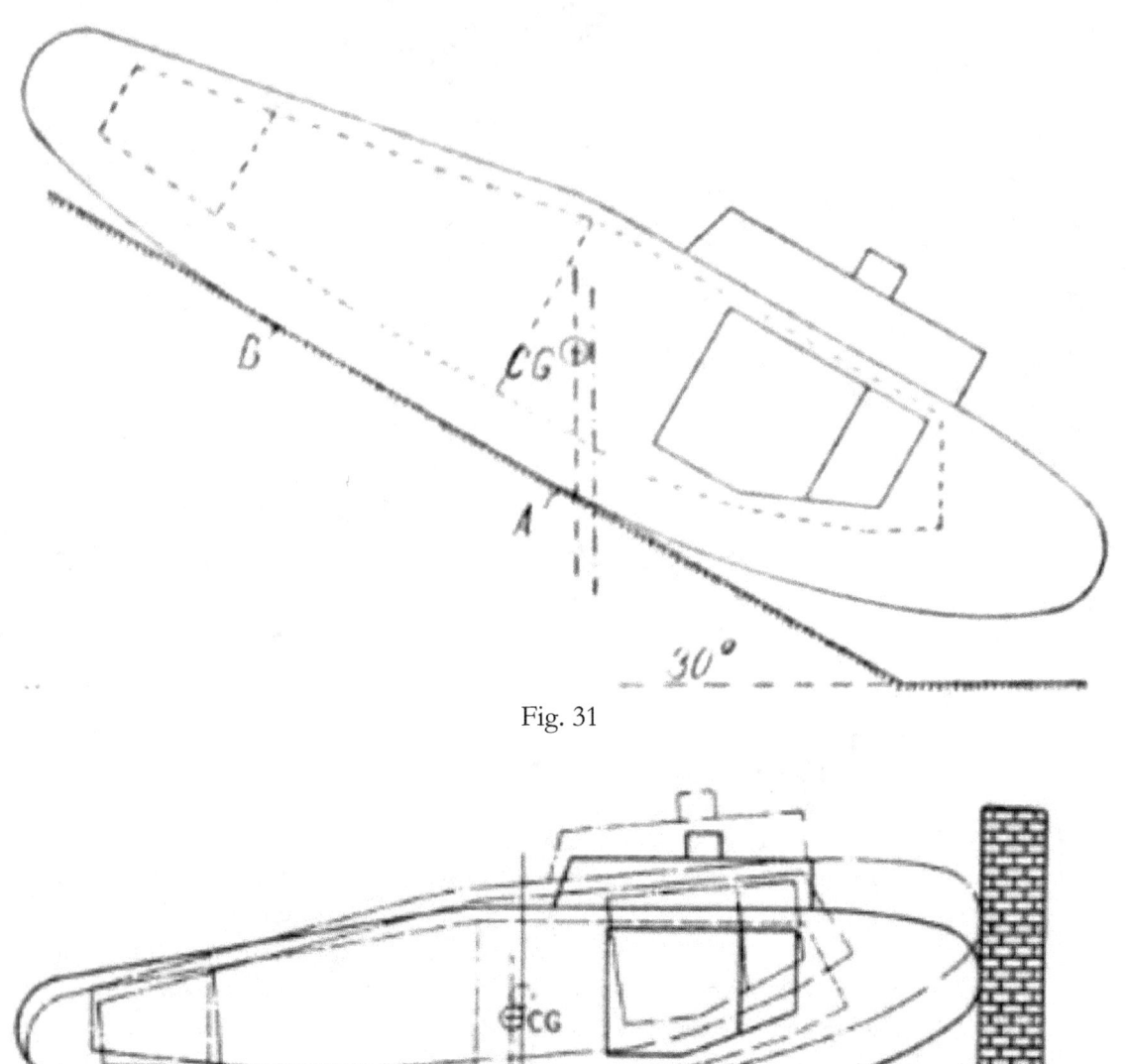

Fig. 31

Fig. 32

POSSIBLE IMPROVEMENTS IN CHAIN TRACKS.

The chain track with steel links wears rapidly on the roads; it runs under extremely bad conditions. In order to obtain overlapping of the links, it has been necessary to place the pin joints with their centres very near to the ground in a position in which no casing can protect them against dust, water or mud. This disadvantage may be somewhat reduced by the abundant use of thick oil as practised [sic: practiced] in America.

Tests of belts made in England in 1915 by Tritton and Wilson did not give the satisfactory results which were expected. This system, however, had already been investigated in Russia by Mr.

Kégresse for the running of cars over snow, and he had attained satisfactory results from belt tracks. Further trials made in February, 1921, in France, showed that the Kégresse flexible belt track enabled a car to be run, not only over snow, but also over sandy ground, even over steep sand, and on marshes.

The hammering of hard and particularly of iron-shod wheels on the road has always been the greatest enemy to high speed in traction. Speed on the road has, moreover, been increased by the use of rubber tires [sic: tyres], and this has been still further increased by the substitution of the pneumatic for the solid tire [sic: tyre].

It is obvious that chain-track traction will find its future to lie in improvements to be obtained from a more yielding contact and more elastic springing than exists at present. If these improvements to be obtained from a more yielding contact and more elastic springing than exists at present. If these improvements follow with the same rapidity as in the case of the automobile, we may in the near future see light cars capable of being run at a high speed on the slopes of mountains, across fields, over snow, and on the sand of the desert, as well as on a good road; for, as General Estienne has said (translated): "The appearance on the battlefield of chain-track mechanical vehicles was an event the importance of which equals that of the invention of gunpowder."

In this paper I have endeavoured to pay just tribute to the engineers whose fertile brains have conceived and elaborated the details, the creators of the tanks and the adaptors of the chain track to carrying artillery, and to express the admiration and thanks of all, particularly to Colonel R. E. B. Crompton, of England, Mr. William Strait, of the United States, to General Estienne, Mr. E. Brillié and to Colonel Rimailho, of France, with the assurance of our profound gratitude.

Before concluding, I wish to express my sincere thanks to the British Army Council, the French Minister of War, and the American Chief of Ordnance, as well as to the Imperial War Museum and the Institution of Mechanical Engineers for the permission which these several Governments and Institutions have been kind enough to give me for the reproduction in lantern slides, illustrations and films of the different machines to which reference has been made.

I wish also to express my sincere thanks to Mr. Brillié, of the firm of Schneider, to Colonel Rimailho and Mr. Berthier, of the Forges et Aciériés de la Marine et d'Homécourt, to Mr. Jannin, of the Renault works; to the Fiat works of Turin, and also to our devoted secretary, Mr. de Dax for the assistance and collaboration which they have all been kind enough to give me in the preparation of this paper.

Tanks and Chain-Track Artillery
Part I.

By L. A. Legros

TWO very different classes of self-propelled vehicles were evolved during the war: first the 'tank,' the armed and armoured machine running on the road or on the ground, crossing trenches, embankments, and entanglements ; and second, chain-track artillery capable of running into position behind the firing line and carrying heavy, long guns or howitzers, fired from the machine.

These machines presented several new problems to the engineer, due to the great weight of the and armament and to the exceptional difficulties of the country to be traversed and the obstacles to surmounted.

Armour

The first practical attempts to use armour for mechanical transport date to the Transvaal War (1899-190.), when Colonel Templar worked out an application of armour to haulage lorries, but these cars proved too heavy for the roads of the Transvaal, where wheeled steam tractors could be used provided that the engine was compound and fitted with a silencer. The steam of the exhaust was not visible in this dry climate, but noise attracted the fire.

It was ascertained during the eighth month of the last war that the protection of any machines and crew would have to be restricted to armour adequate for stopping the German rifle bullet - that is, from 0.40 in. to 0.43 in. thick; but a little later it was found that this
could be pierced by the reversed bullet fired at close range, and it was necessary to increase the thickness progressively up to 0.63 in. and even more for the front and sides of the machines. In the earlier and in the lighter types of tank the thickness of the back armour ranged from 0.25 in. to 0.33 in., and in the latest it the same as the front and sides. The thickness of the roof varied from 0.2 in. to 0.3 in.; 6 mm. being very general.

The experience gained with armoured cars was practically the branch of automobile engineering that could afford any data on which the Consulting Engineers to the Admiralty Landship Committee, Colonel Crompton and the author, could commence the work of designing an armoured trench crossing vehicle to meet the conditions laid down by Mr. Winston Churchill, Chief Lord of the British Admiralty.

* Abstract of a paper read before the Societe des Ingenieurs Civils de France on July 8th, 1921, in Paris, and before the British Section of the I.C.F. (in English) on October 19th 1921.
Particulars of commercial and agricultural tractors will be found in a paper by the author, 'Traction on Bad Roads or Land', read before the Institute of Mechanical Engineers in January 1918 (Proc. I.Mech.E., 1918, pp.55-194). The British tanks are very fully described in the paper read by Sir

Eustace Tennyson d'Eyncourt, K.C.B., before the British Association on September 10th 1919, and reprinted in Engineering, September 12th and 19th, 1919.

Details of some other features of the experimental work and of problems connected with the tanks are given in an abstract of the present paper in the Engineer, December 2nd, 9th, and 16th, 1921.

The chain-track.

It became evident at an early date that the wheeled underframe or chassis could not carry the great combined weight of armour, armament, and propelling machinery, and that another method of distributing the load and propelling the machine across country, particularly over the soft ground of Flanders, would be necessary. The Wheel on hard ground can carry a load up to 30 lb. per sq. in. with a sinkage of 0.2 in. to 0.3 in.; the insistent load can increased by enlarging the diameter or the width of tread of the wheels, but only within narrow limits on account of the difficulties of transport of the tractor by rail.

The chain-track or self-laying railway invented by R. L. Edgeworth in 1770 had been revived in various forms, such as the Boydell wheel of 1854 and the girdle, still used on the wheels Of German the War. As a self-laying railway it was adopted in the United States in 1904 in the Lombard tractor, which used two separate superposed chains, the outer forming the chain-track proper, and the inner, fitted with rollers, running between the surface provided on the outer chain and a fixed roller-path carried on the tractor (fig. 1). This arrangement of intermediate roller-chain, much improved, was adopted later by Diplock in his last models of the Pedrail. With this system the resistance to traction may be from 50 to 60 lb. per ton as compared with 140 to 230 lb. ton with the single chain track, under good conditions giving comparative immunity from dirt and clay ; such conditions existed in the log-hauler, a machine built to haul logs or timber over snow, and in the Pedrail of 1915 , in which the roller chain was carried some 5 or 6in. above the ground.

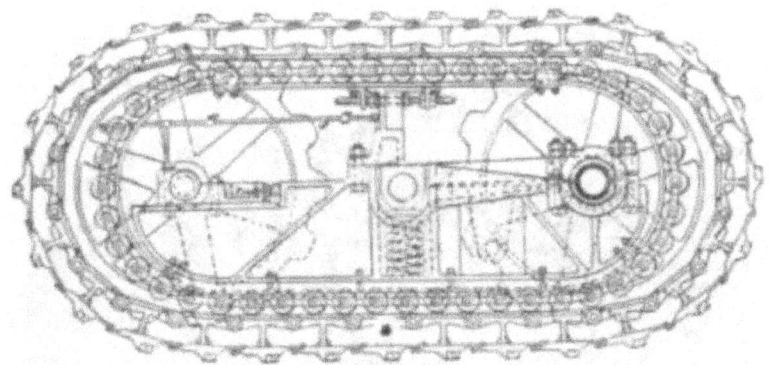

Fig. 1. The Lombard track.

In 1904, while the Lombard machines were commencing to work at lumbering in America, Roberts in England, was making experiments with a Darracq car and also with another vehicle fitted with a paraffin-engine (fig. 2). Roberts used a single chain which he constructed so as to form an inverted *voussoir* (arch) and thus gun increased the diameter of the arc of contact with the ground; he obtained, as a supporting surface, an arc of a wheel 33 ft. in diameter with a total height of vehicle of less than 9 feet. These cars could be run over soft sand, and even marsh which a horse could not cross. The Roberts machine was tested by the War Office, and, apart from the difficulty of digging itself in when turned in a small circle, it was found to be liable to pick up stones between the radial surfaces of the links, producing a 'nut-cracker' action capable of causing breakage, jamming, or the derailing of the carrying wheels of the vehicle from the chain.

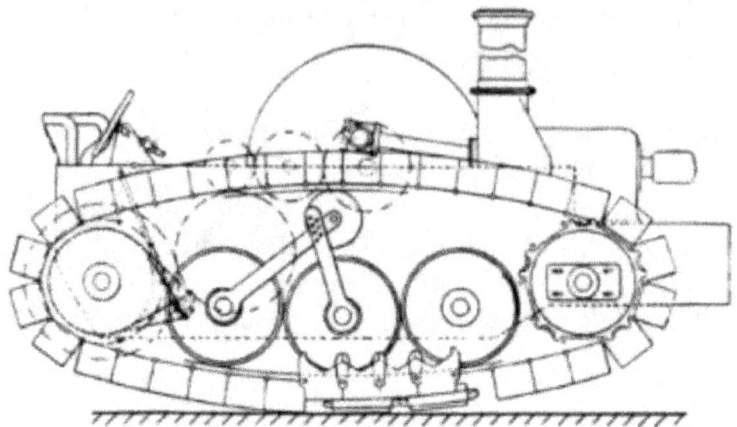

Fig. 2. The Roberts endless-track tractor with paraffin engine.

The Roberts tractor was known officially in England as the Roberts endless track tractor, and unofficially under the name of the 'Caterpillar.' One of these vehicles, under the charge of Major Donohue , C.B.E., then Chief Inspector of Military Mechanical Transport, was run in demonstration in 1908, at Aldershot, before King Edward VII. The practicability of carrying a cannon on chain-tracks was afterwards shown on May 18th, 1908, by means of a wooden, full-size, mock-up constructed by Major Donohue* (fig. 3).

Fig. 3. Major Dononhue's mock-up of chain-track gun carriage.

Although the chain-track found little favour in England, its development continued on commercial lines in America, with improvements in springing and articulation of the trucks, by Holt, who adopted the name 'Caterpillar' in 1905. Advantage was taken of English experience, and the chain details were altered so that the pin-joints were placed near the ground and covered by part of the links in such a way that no opening was left between the links when passing over the driving pinion or guide-pulley and the nut-cracker difficulty was eliminated, but at the expense of placing the working joints nearer to the ground level (fig. 4).

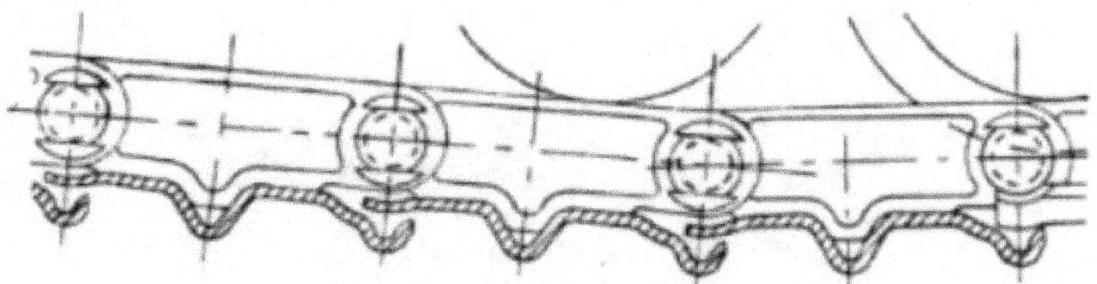

Fig. 4. Chain-track of British tanks.

Among the American improvements, the invention by W. Strait (December, 1912) of the climbing chain-track, having the climbing part inclined to the carrying part, was destined to have a very great influence in the future. This arrangement is shown diagrammatically in fig. 5. Strait describes his invention as follows:-

"The ground stretch is formed with a bulge or peak (A) between its two ends, and an inclined part between (B) and (A) and a similar inclined part between (A) and (C). If a large stone, stump, or other obstruction is encountered it will easily pass under the forward sprocket, the forward incline will ride up the obstruction gradually until the peak (A) is reached, and then the rear incline will gradually ride down the obstruction."

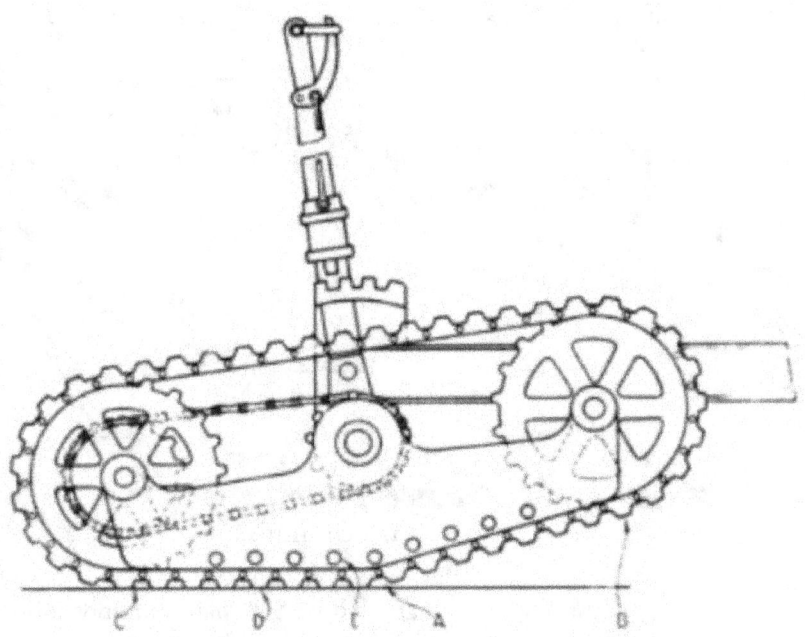

Fig. 5. Strait's climbing chain-track.

A Nash Quad tractor of American make, fitted with a track of this form is shown in fig. 6. The form of the track was somewhat modified in the actual Killen-Strait tractors built in 1914, which also had a triangular track, but not of isosceles form (fig. 7). In this tractor the further feature was added of a separate carrying track in the front for steering and carrying part of the load (fig. 8).

Fig. 6. Nash Quad fitted with Strait pattern climbing track.

For Crossing trenches and embankments of earth the climbing caterpillar was not necessary, and it will seen that the Schneider, St. Chamond, and some other tanks did not embody this principle, the formation of the front, like the prow of a boat, proving adequate for crossing obstacles of earth.

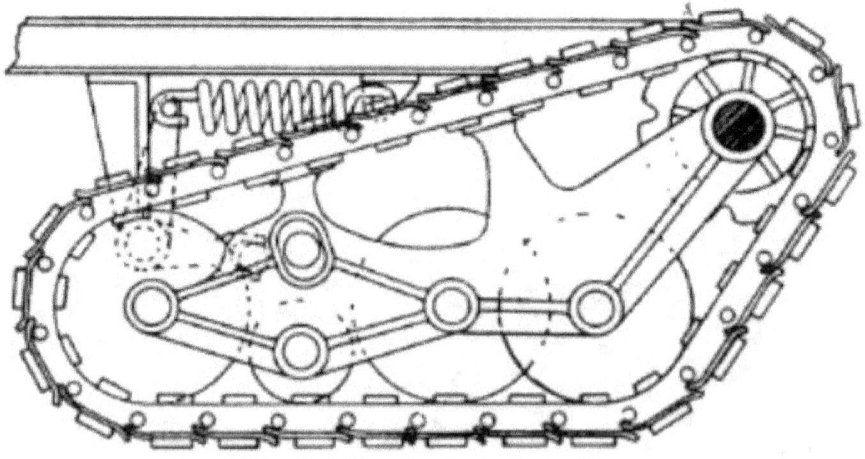

Fig. 7. Killen-Strait chain-track.

The use of a front carrying-track was suggested by St. Chamond, independently of Strait, in but the scheme was not executed.

Fig. 8. Killen-Strait tractor with leading carrying chain-track.

Steering.

Several methods were adopted by the makers of chain-track tractors; the use of leading wheels with the ordinary Ackermann axle being adopted by the Creeping-Grip and Allis-Chalmers ; a single carried in a frame by Holt, Yuba and the Tracklayer ; a leading chain-track carried in a frame by Strait ; and the arrangement of Roberts, that of the whole load on the two tracks, was being adopted before the war by Holt and Creeping-Grip for their small powered 'Baby' tractors, the former having already constructed a model and the latter having one in progress. Steering with this arrangement involved the declutching of the inner track in turning, with or without the supplementary action of brakes as in the Holt, or a reversing gear as in the Creeping-Grip ; the latter arrangement permitted the vehicle to be turned on its axis instead of pivoting about the centre of one of the track-chains.

Early experiments.

The firm of Schneider et Cie., in France, obtained two Holt caterpillars, one being of the 'Baby' type, and these machines were used for tests at Le Creusot in May, 1915; with the superiority of the 'Baby' type, without steering wheels, for handling on rough ground, was early [sic: clearly] apparent, and demonstrations were made in the presence of the President of the French Republic on June 16th, 1915. In July the construction of an armoured chain-track machine was commenced. Further trials and demonstrations were made on the uneven ground at the Front at Souain on December 9th, and at Satory camp On December 27th, 1915. Colonel (now General) Estienne, having seen the work done by the Holt tractors on the British Front, suggested the production of landships to act with infantry offensive operations. On December 10th, 1915 he was placed in communication with the Schneider works; a new design was put in hand in accordance with the particulars he furnished, and the scheme the Schneider Estienne *char d'assauat* (or tank) was communicated to the French G.Q.G. (G.H.Q.) on December 27th, 1915.

The possibility of steering with a greater length of chain-track had not yet, however, been ascertained, and trials were made on February 21st, 1916, at the Vincennes Polygon, which showed the suitability of the proportions proposed by the constructors. On February 25th, 1916, an order was given to Schneider et Cie. for 400 of these tanks (figs. 9 and 10) in place of the 10 motor machine-guns on which they were engaged, the designs and work being entrusted to our distinguished colleague, M. Eugène Brillié.

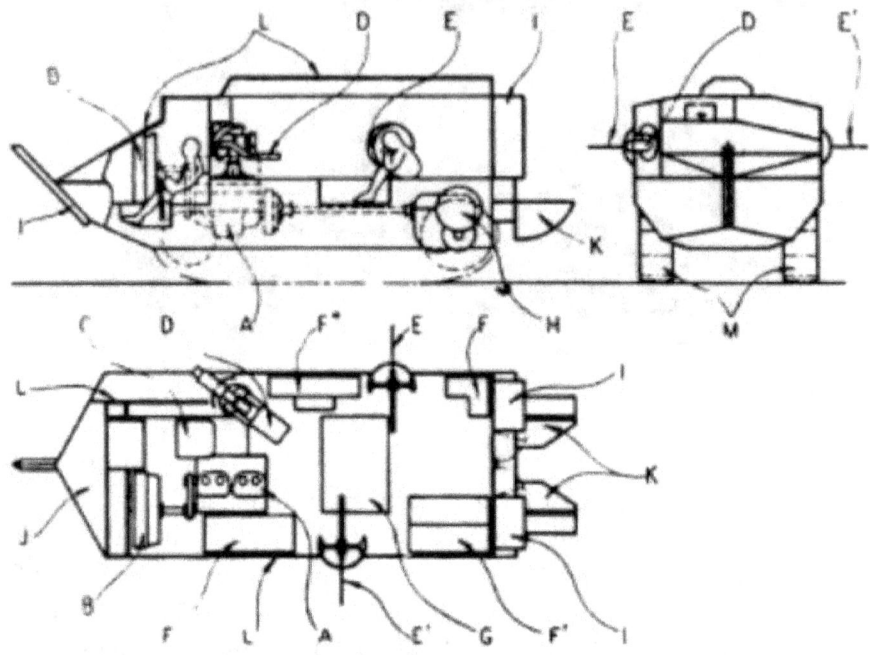

Fig. 9. The Schneider char d'assaut, sections.

A – Engine, B – Radiator, C – Driver's seat, D – 75 mm short gun, E – machine guns, F – munitions, G – Well in floor, H – Reduction gear, I – Petrol tanks, J – Prow, K – Tail wings, L – Armouring, M – Chain-tracks.

Fig. 10 The Schneider *char d'assaut*.

The roughly constructed machine that was used for the trials of February, 1916, served also as the basis for another pattern of tank, the design of which was undertaken by the Forges et Aciéries de la Marine et d'Homécourt, usually known as St. Chamond, and resulted in 400 tanks Being built by this firm (fig. 11).

Fig. 11. The St. Chamond char d'assaut.

This tank was fitted with a motor of 90 h.p. running at 1,450 revolutions per minute, of the four-cylinder, sleeve-valve Panhard type. The total weight of this tank in running order, without munitions or crew, was 19.9 tons, and with munitions and crew 21.5 tons. One of the most important characteristics of this vehicle consisted in the electric transmission with infinitely variable change of speed, which avoided the great difficulty in changing gear of the early English tanks.

The St. Chamond tank carried a 75 mm. quick-firing St. Chamond gun, or an 1897 pattern 75 mm. gun, with its supply of cartridges. The contents of the petrol tanks were 265 litres (58 gallons), distributed among three petrol tanks; this supply was sufficient for running a distance of 35 km. (22 miles) on slow speed, and of 60 km. (37 miles) on top speed.

Steering was effected by means of a steering handwheel acting on contactors, the closing or Opening of which enabled different combinations of the internal connections of the electric motors to be made or broken. These combinations acted by accelerating the outer motor and slowing, braking, or locking the inner motor on the curve.

In England, Mr. Winston Churchill, as already stated, formed a Landship Committee of the Admiralty to work out various schemes, and a grant of £80,000 was assured by the British Treasury. It was not possible to obtain 'Baby' Holt tractors for the necessary experiments, they were required by the War Office: the 'Baby' Creping-Grip was not yet complete or tested. As a basis for experiment two 'Giant' Creeping-Grip tractors, and one Killen-Strait tractor were ordered from America, while attempts were made to adapt the only English chain-track, the Diplock Pedrail, to the end in view. These Diplock chains and chain-tracks were subsequently used on an experimental flame-throwing machine.

No particulars were available in England at this time as to the amount of power absorbed in driving the chain-track, the ratio of draw-bar pull to adhesive weight, or the insistent loads that were either usual in existing American machines or necessary for the conditions at the front. Permission for the consulting engineers to visit the front was refused, and the early English experiments had to be carried out on ground and trenches prepared in accordance with the usual military practice of the time.

Nevertheless, experiments made with the Killen-Strait tractor early in June, 1915, terminating with a demonstration before Mr. Lloyd George at the end of the month, showed that the chain-track tractor was capable of crossing the obstacles expected to be met and of crushing barbed-wire entanglements.

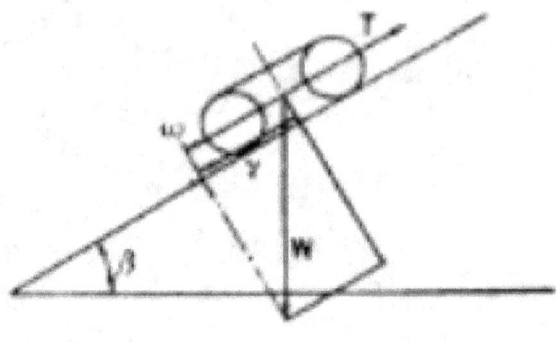

Fig. 12.

These experiments were repeated on a larger scale early in August at Burton-on-Trent with entanglements of heavy construction, reinforced with steel wire rope over half an inch in diameter. At the end of July, 1915, Colonel (now General) E. D. Swinton, C.B., D.S.O., obtained particulars semi-officially, which gave a better idea of the actual difficulties to be overcome. The Killen-Strait tractor during the trials that it could be driven even when the front carrying track was of the ground, and it demonstrated, when run backwards, the peculiar utility of the inclined portion of the track in climbing obstacles. The experiments afforded inspiration to many inventors, among whom may be mentioned Messrs. Macfie, Wilson, Tritton, and Nesfield.

During 1915 the French and English engineers were, unknown to each other, working on very similar lines the seal of secrecy. The difficulty of steering with very long tracks could not be considered as solved by the 'Baby' Creeping-Grip with a base of only 4 ft., or by the 'Baby' Holt with a base of 5 ft. 4 in., and with the object of making full-size trials to settle this doubtful matter the consulting engineers ordered in May, 1915, two of the Creeping-Grip track; lengthened to give a 9 ft. base in contact with the ground.

The first English designs, commenced in February, 1915, were for a machine intended to carry a storming party of seventy men, but it was soon ascertained that the length was too great for negotiating the curves and turnings of the French and Belgian roads; it was to divide the landship into two parts in a manner somewhat similar to that subsequently and successfully adopted by St. Chamond in France. The Pedrail was abandoned and ordinary chain-tracks Were adopted. Instructions were then received that the designs were to be altered to offensive machines carrying two gun-turrets, and the order was subsequently reduced to one turret only with the object of keeping down the height. The experiments made with the American tractors had given much of the necessary information required for a sound design; it was ascertained that a drawbar-pull of about 60% of the weight could be obtained; hence it was that the machine could climb a gradient of about 40 degrees. In some cases the tractive effort on tanks and chain-track artillery has amounted to as much as 80% of the weight of the vehicle.

Gradient Climbing.

The calculation of the gradient that can be climbed by a tractor is obtained as follows (see fig. 12):

T represents the tractive effort,
W the weight of the vehicle or train of vehicles,
R the coefficient of resistance to rolling (as a fraction),
w the component of W parallel to the ground,
r the resistance to rolling,
$ß$ the gradient measured as an angle ;

$$T = r + w = W(R \cos ß + \sin ß)$$

when $ß > 6°$ (say in 1 in 10) it is usual to assume that $\cos ß = 1$. Then, when n is the gradient measured as a percentage.

$$T = W(R = n/100)$$

This, while applicable to the conditions of railways or easy roads, cannot be adopted for the heavy gradients climbed by tanks for which the normal formula must be used.

A convenient construction for determining the value of T by graphical methods, due to M. Brillié, is shown in fig. 13. If $ß$ *be* the angle of the gradient, set off AC to represent W and CD at right angles to it to represent RW. On AC describe a semicircle, cutting AB (the line of gradient) in B, and on CD describe a semi-circle. Join BC and produce it to cut this semi-circle in E; then BE will represent the tractive effort required.

TABLE I.—PARTICULARS OF TANKS CONSTRUCTED DURING THE WAR.

Country.	Tank.	Tanks built (approx.).	Valve gear or make.	No. of cylinders (per eng.).	Bore. mm.	Stroke. mm.	Standard speed, revs. per min.	H.p. at standard speed.	Weight fully equipped. Tons.	H.p. per ton (fully eq.).	Length. ft.	Maximum width. ft.	Maximum height (h). ft.	Ground clearance, no sinkage. in.	Width of chain-tracks. in.	Length of track on ground, no sinkage. in.	Insistent load for 2 in. sinkage. lb. per sq. in.	Insistent load for 4 in. sinkage. lb. per sq. in.	Crew—men.	Fuel, gallons (Brit.). Gal.	Radius of action, miles. Miles.	Height of front guide wheel axis, in. in.	Carrying wheel axles per track.	Slow speed, m. per hr. Miles per hr.	Top speed, m. per hr. (V). Miles per hr.	Gear change: s=sliding; ep=epicyclic; el=electric.	Tractive effort at track (efficiency 75%); tons. Tons.	Immunity factor V^2/h.
Britain	Mark I (wheels)	150	Sleeve	6	150	150	1,000	105	27.5	3.82	32.5	13.8	7.4	16.5	20.5	55	24.6	16.9	8	46	23	61	26	0.73	3.70	s.	17.8	1.85
,,	Mark II [III, IV]	1,115	Sleeve	6	150	150	1,000	105	27.5	3.82	26.4	13.8	7.4	16.5	20.5	55	24.6	16.9	8	46	23	61	26	0.75	3.70	s.	17.8	1.85
,,	Mark V [VI, VII]	1,036	Ricardo	6	143	190.5	1,200	150	28.5	5.26	26.4	13.5	7.4	16.5	20.5	55	19.4	13.2	8	93	45	61	26	0.90	4.00	ep.	20.5	2.86
,,	Mark IX	35	Ricardo	6	143	8.1	1,200	150	37	4.05	31.9	21.2	7.7	21.2	20.5	72	24.0	19.1	8	100	42	50	33	0.86	4.30	ep.	22.0	2.40
,,	Mark A [B]	245	2 Tylor	4	127	152	1,000	90	14	6.43	20.9	8.6	8.7	22.0	20.5	48	14.9	13.6	3	70	80	34	16	1.30	8.30	ep.	9.3	7.92
France	Schneider	400	Schneider	4	125	170	1,000	60	14	4.29	20.7	6.7	7.6	15.8	19.7	75	11.0	8.5	7	77	47	22	7	1.24	4.16	s.	10.5	2.28
,,	St. Chamond	400	Sleeve	4	125	150	1,350	85	21.5	3.95	26.0	8.8	7.8	19.7	19.7	104	10.0	9.0	8	58	37	20	7	0.75	10.50	el.	10.1	3.21
,,	Renault	2,680	Poppet	4	95	160	1,500	39	6.7	5.82	13.5	5.7	7.0	15.8	13.0	81	6.3	5.7	2	20	28	26	7	0.93	4.85	s.	5.0	3.36
Ang.-Amer.	Mark VIII Int'l.	108	Liberty	12	127	178	1,400	330	37	8.92	34.2	12.3	9.4	21.0	26.5	90	17.4	16.0	8	200	55	60	29	1.40	5.20	ep.	31.5	2.88
American	6-ton	950	Buda	4	108	140	1,200	40	6.6	6.08	15.6	5.7	7.6	16.0	13.4	64	6.0	5.3	2	25	30	22	6	0.88	4.52	s.	5.8	2.69
,,	3-ton	15	2 Ford	4	95	102	1,700	34	3.0	11.18	13.8	5.3	5.4	12.0	8.0	50	7.1	5.8	2	12	—	29	6	2.80	11.10	ep.	1.5	11.10
Italy	Fiat 2,000	—	Fiat	6	160	180	1,200	160	42	4.75	24.3	5.9	12.5	21.3	17.7	85	22.1	18.1	10	132	47	41	8	0.62	4.66	s.	40.5	1.74
,,	Fiat 3,000	—	Fiat	4	105	180	1,500	45	5.5	8.20	13.8	5.4	7.2	13.8	11.0	78	6.7	5.6	2	20	112	20	8	1.24	9.94	s.	4.0	1.72
Total		7,134																										
Germany		15	2 Daim.	4	165	200	800	200	40	5.0	40.0	10.5	11.0	9.0	20.5	174	11.8	11.2	16	—	—	75	15	—	8.11	s.	—	5.98

193

TABLE II.—DETAILS OF FRENCH AND AMERICAN CHAIN-TRACK ARTILLERY.

Country.	Self-propelled artillery.	Bore of gun. mm.	Type of gun.	Weight of gun. Tons.	Weight of projectile. lb	Maximum range. Yards.	Maximum elevation. Deg.	Engine.	Standard speed, revs. per min. r.p.m	H.p at standard speed. H.p.	Weight fully equipped. Tons.	H.p. per ton fully equipped. H.p. per ton	Length. ft.	Maximum width. ft.	Maximum height. ft.	Ground clearance, no sinkage. in.	Width of chain tracks. in.	Length of track on ground, no sinkage. in.	Insistent load for no sinkage.	Insistent load for 3 in. sinkage.	Insistent load for 6 in. sinkage.	Fuel, gallons (Brit.).	Radius of action, miles.	Carrying wheel axles per track.	Slow speed, miles per hr.	Top speed, miles per hr.	Gear change: s=sliding; ep=epicyclic; el=electric.
France...	Schneider automotor gun carriage	220	L	14·0	221	24,000	37°	Schneider	1,200	150	40·0	3·75	25·9	9·0	9·6	8·1	25·5	169	—	10·5	8·5	—	—	17	0·62	4·41	s
,,	St. Chamond chain-track gun carriage	194	F	8·6	176	20,800	35°	*	1,200	120	29·6	4·05†	27·9	8·5	9·8	19·7	—	110	—	12·4	—	88	—	13	2·49	4·97	el.
,,	St. Chamond chain-track gun carriage	280	S	4·1	447	12,000	60°	*	—	120	29·5	4·07†	24·6	8·5	9·8	19·7	19·7	110	—	12·2	—	88	—	13	2·49	4·97	el.
America	Mark I 8-in. howitzer ...	208	—	3·5	200	12,100	45°	Art. Trac. Duesen- berg	850	80	26·3	3·04	23·7	9·0	9·0	—	18·0	117	—	—	—	29	—	10	0·93	4·00	s:ep.
,,	Mark III mortar	240	—	4·9	336	16,500	60°	*	1,200	225	49·5	4·55	24·8	9·8	9·4	—	24·9	152	—	11·5	—	83	—	15	0·87	4·30	s:ep.
,,	Mark VII 75 mm.; 1916 model	75	—	0·34	16	9,700	45°	Cadillac	1,500	35	4·8	7·28	11·3	5·2	5·9	—	[8·0]	50	—	7·3	—	—	—	6	3·00	9·30	s
,,	Christie (wheeled) gun-carrier	155	L	3·97	93	17,200	35°	Christie	1,200	120	20·0	6·00	19·7	9·3	6·6	12·6	22·0	165	6·1	—	—	33	34	4	1·75	14·7	s.

* Electric drive from leading vehicle carrying ammunition. † Gun-carrier alone running.

194

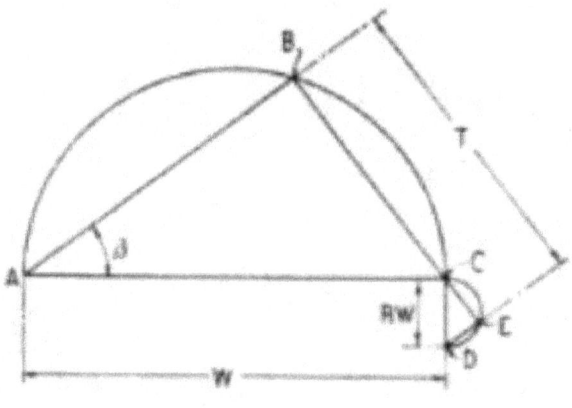

Fig. 13.

If T be given and it desired to find the gradient that can climbed, the scale is then placed the drawing so that its edge kept on C while its end is on one of the semi-circles. While observing these conditions, the scale is rotated about C till the division corresponding to T coincides in B (or E) with the other semi-circle; then the angle BAC is that of the gradient that can be climbed.

The double vehicle which had been found a very successful combination in crossing difficult country in South Africa and also in South America was, however, not adopted in England, though it was proved successful later in France by St. Chamond for their heavy gun carriers.

In England the lozenge form of tank with the enveloping chain of Wilson and Tritton was adopted, and gave ability to cross wide trenches and to climb obstacles of reinforced concrete. These early tanks, however, had certain disadvantages due to the multiple control and the difficulty of gear changing by sliding gears in a very slow machine. Later patterns were fitted with one-man control and epicyclic gear change.

The essential figures in the schemes of the clouting engineers were 5.5 h.p. per ton, 4.5 miles hour and an insistent weight of 8 lb. per sq. in. Table I. shows the gradual increase made in England from 3.8 to 11.2 h.p. per ton, from 3.7 to 7.75 miles per hour and the reduction from 24.6 to 15.0 lb. per sq. in. insistent weight with a 2 in. sinkage. It may also be noted that the Renault tank designed on the basis of French experience early in 1917 had 5.8 h.p. per ton, a top speed of 4.85 miles per hour and 6.3 lb. per sq. in. insistent load with 2 in. sinkage.

As soon as the English tanks and the Schneider and St. Chamond *chars d'assaut* had shown the capabilities of this new arm, various other designs were put in hand, of which the most noteworthy was probably the Renault 6-ton tank (fig. 14), of which pattern more were built by this firm than the total number of all marks constructed in England during the war.

Fig. 14. Renault *char d'assuat*.

The method of springing the chain-track frames consisted of plate springs, compensating levers and bogies, which ensured a sensibly constant and uniformly distributed load over each of the carrying rollers whatever might be the irregularity of the ground to which the chain must adapt itself. The tank was controlled entirely by man. The turret carried on a ball-race could be swung round its vertical axis so that the machine gun or quick-firing gun would cover the whole of the horizon. The Renault works were turning out these tanks at the rate of twelve per day before the Armistice was signed.

Tanks and Chain-Track Artillery
Part II.

By L. A. Legros

The American tanks.

WHEN America entered the War very little authentic information covering the use and types of tanks was available in that country. Some rather vague specifications were obtained from France during the summer of 1917, from which two patterns of experimental tanks were built. One of these was steam driven, the other equipped with a petrol-electric drive. Before the experiments with these had been concluded, the great importance of the tanks had made itself apparent. The 'Ordnance Department' of the United States then sent one of its officers to Europe cm a special mission with the object of securing all available formation relating to the construction and use of the tanks. After numerous conferences with the British and French authorities, America undertook the construction of two types of tanks for the American Army the small two-man tank based on the French Renault tanks, and the large 30-ton tank to be produced jointly with England.

Renault tanks were purchased and sent to America with a complete set of drawings, and instructions were given to duplicate these tanks, using American standards of manufacture and measurement.

The designs and specifications relating to the large Anglo-American Mark VIII tank were completed after numerous conferences with the British General Staff, and the orders were then put in hand for simultaneous manufacture in England and America.

According to a treaty signed by the two countries on January 22nd, the Commissioners appointed under this treaty were empowered to the designs, to arrange for the production of British and American Components, to build a factory in France for the assembly of these components and actually to assemble the tanks.

The designs were commenced in November, 1917, and by May, 1918, practically all the drawings had been received in America, where the production of parts, including motors, transmission, radiators, and chain-track rollers and sprockets, was to take place. The British Government to supply the chain-tracks, armour, framing, and armament; the British works had greater experience in the production of armour-plate and possessed a large stock of guns and other fittings necessary for the armament of the tanks.

During this period the construction of American tanks of the Renault type, figs. 15 and 16, proceeded, and the assembly was entrusted to three large American works: the Van Dorn Iron Works, Cleveland (Ohio), the Maxwell Motor Company, Dayton (Ohio), and the C. L. Best Companv, also of Dayton. The order was for 4,440 tanks, of which only 950 were actually completed.

Fig. 15. American 6-ton tank.

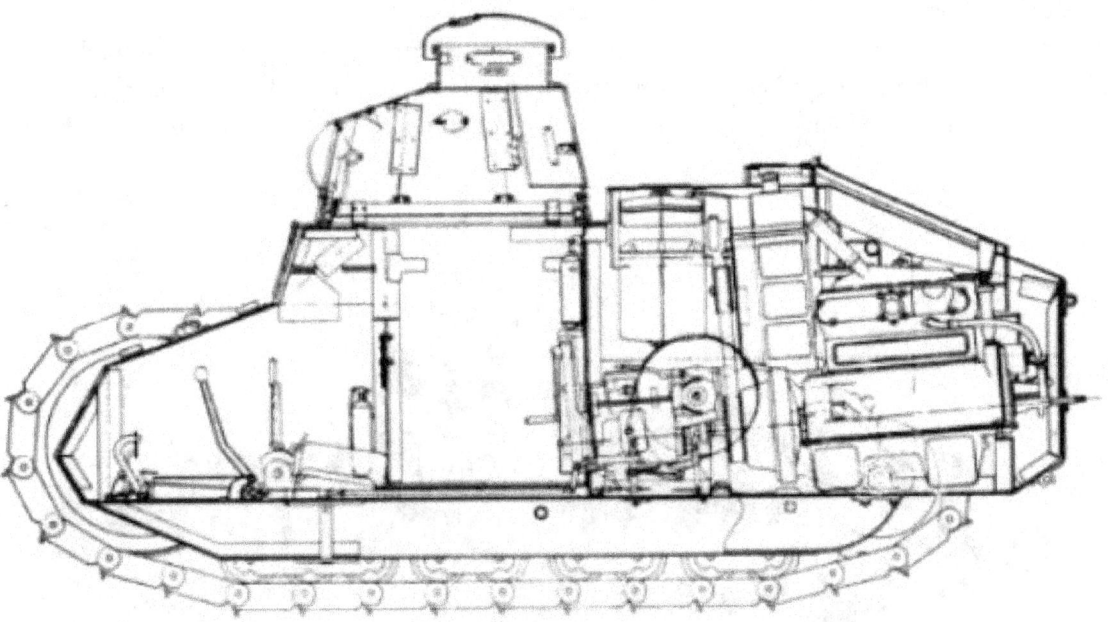

Fig. 16. American 6-ton tank.

During the Summer of 1918 work was continued on the Mark VIII tanks, fig. 17. A sample set of British components was sent to America, and with the addition of the American components, which had been quickly prepared, the assembly was carried out of a complete tank which was tested at the works of the Locomobile Company, Bridgeport, Conn. Everything was in good order in England and in the United States, as well as at the American works at Neuvy-Pailloux, in France, which was ready for full production when the Armistice was signed. The manufacture of 1,500 tanks was then in hand with a view to a heavy delivery in the spring of 1919.

Fig. 17. Anglo-American Mark VIII tank.

America, moreover, had undertaken the construction of 1,450 complete Mark VIII tanks from parts manufactured entirely in the United States, in addition to the components which were already being made there.

During the summer and autumn of 1918 a 3-ton tank (fig. 18), much smaller than the Renault, was designed by the 'Ordnance Department' to be constructed by the Ford Motor Company of Detroit. These designs made use as far as possible, of the Standard parts of the Ford automobile.

Just before the signing of the Armistice orders were placed for 15,000 of these 3-ton tank (fig. 19), which were intended to be used either as light tractors or as machinegun tanks, America had completed one hundred Mark VIII tanks in the spring of 1920 for the service of the 'American Tank Corps.'

Fig. 18. American experimental 3-ton tank.

Fig. 19. American 3-ton tank.

The Italian tanks.

By the order and at the expense of the Italian Government designs for the construction of tanks were undertaken by the Fiat firm at Turin. The first tank (fig. 20) was of heavy pattern, weighing 35 tons, with a 250 h.p. motor and a maximum speed of 7.4 miles per hour. The armament consisted of a short gun carried in a revolving turret.

The Fiat firm has recently produced a lighter tank (fig. 21), the weight of which is about 6 tons. The motor is of 45 h.p., running at a normal speed of 1,500 revs. per minute, and giving a maximum speed of 16 km. (1o miles) per hour; the armament consists of two coupled machine guns arranged in a revolving turret.

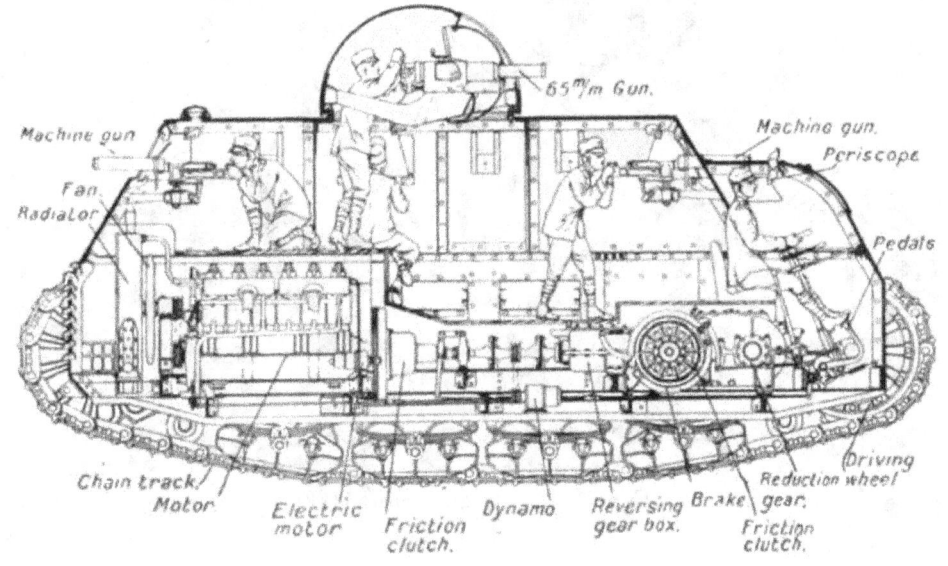

Fig. 20. Italian tank, type 2,000.

The German tanks.

The Germans produced only one model of tank, a large, heavy machine of which only fifteen are believed to have been actually constructed.

Skeleton instruction tanks.

Not only was it necessary to produce large quantities of the actual fighting machines, but special machines of skeleton construction were necessary for training the allied crews. An example of an American skeleton tank is shown in fig. 22.

Auxiliary motor tractors and lorries.

In order to keep the tanks supplied with munitions and stores, tractors were required capable of crossing the same country and climbing the same gradients. The British tanks, with their high insistent weight, did not demonstrate this necessity so clearly as did the French *chars d'assaut*, in which this factor was much lower. The firm of Schneider et Cie. produced a chain-track lorry known as CD type (fig. 23) with the same chain-tracks and two-truck springing arrangement as used on their *chars d'assaut*, capable of carrying a load of 4 tons and of giving a drawbar pull of from 7 to 8 tons.

The engine and the greater portion of the driving mechanism were the same as in the Schneider *char d'assaut*. The weight, light and in running order, was 10 tons, there were four forward speeds from 1.3 to 5.15 miles per hour, and the same in reverse; steering was effected by declutching and braking the inner track; a winch driven by engagement with the first speed gear was fitted for hauling; the petrol tank capacity was 44 British gallons, and the tractor with a 3-ton load could climb a gradient of 50 per cent. on the first speed at a rate of 1.3 miles per hour. The insistent weight for 2 in. sinkage with a 4-ton load was the same as that Of the Schneider *char d'assaut*.

Fig. 21. Italian tank, type 3,000

Fig. 22. American skeleton tank.

Fig. 23. Schneider ammunition chain-track tractor, type CD2.

Another variety of tractor, known as type CD3 (figs. 24 and 25), was produced by Schneider et Cie. for carrying the 155 mm. long and short field guns, and the 220 mm. howitzers complete with their carriages across bad country. This tractor had a similar arrangement of chain-track to the CD type, but longer, and the width of the tracks was increased from 350 mm. to 450 mm. (from 13.75 in. to 17.75 in.).

Fig. 24. Schneider gun-carrying tractor; preparatory to loading.

Fig. 25. Schneider gun-carrying tractor; gun secured for transport.

The chassis frame was of special construction, carrying a small rotating jib crane used for lifting the end of the gun carriage to its initial position for hauling into place : a fixed jib which projected over the radiator carried a sliding block, a projection on which, engaging with the end of the carriage, enabled the gun to hauled into place, the wheels of the gun-carriage running up two detachable ramps till in the carrying position, when it was secured the carriage and the wheels. This tractor itself, running order, weighed 12 tons ; it had an overall length of 21 ft., an overall width of 7.9 ft., and a ground clearance of 16.5 in. ; the overall height was 9.8 ft. The engine was the same as in the other Schneider tractors; there were four speeds forward and reverse ranging from 0.7 to 4 miles per hour. The weight of the 155 mm. short gun was 3.3 tons; that of the 155 mm. long gun was 8.8 tons. Loaded with this the total weight was 20.8 tons. and the insistent weight was from 9.5 lb. to 11.5 lb. per sq. inch according to the sinkage.

A heavy-gun chain-track platform-lorry (fig. 26) was also made in quantity by Renault.

Fig. 26. Renault gun-carrying platform tractor.

Chain—track artillery.

In 1915 and 1916 Colonel Crompton and the author prepared designs for a chain-track motor carrying a 4.5 in. howitzer; this machine was designed to use two Lanchester engines, one for each track, with steering by varying the engine speeds - a system adopted later in the Whippets.

Gun-carrying tanks were made later to the designs of Majors Wilson and Greg, forty-eight in all being produced. The idea was, however, carried out much more thoroughly in France, where Schneider et Cie. constructed a self-propelled chain-track mounting for a 220 mm. (long type) gun. The gun, which could be fired at a maximum angle of 37 degrees to the horizontal (fig. 27), was carried on its small normal type of mounting resting on two inclined roller-paths forming part the chassis of the vehicle; a hydraulic brake was interposed between the small mounting and the chassis, and the return to loading position was effected by gravity.

Fig. 27. Schneider chain-track motor mounting, with 140 mm, long type.

The motor was 6-cylinder of 120 h.p.; total weight, 40 tons. The gun was laid roughly for direction by means of the chain-tracks, traversed by the main motor, and the fine adjustment effected either a handwheel or by a small auxiliary 10 h.p. motor driving through a high-ratio reduction gear.

The St. Chamond firm designed and built a carrying tractor for the 120 mm. (long type) quick-firing gun, as well as a motor chain-track gun-mounting carrying a 120 mm. (long type) long-range gun.

The Saint-Chamond chain-track artillery, consisting of two chain-track vehicles per unit, is particularly worth attention; two types were made, the one with a 194 mm. Mark F quick-firing gun (fig. 28) (trailer), and the other carrying a 200 mm. short howitzer.

The unit in each case consisted of two chain-track vehicles (fig. 28); the forward motor - track vehicle, fitted with an electric generating set, carried the ammunition; the chain-track gun-mounting was fitted with two electric motors and supplied from the generating plant on the forward vehicle by means of a cable 50 m. (55 yards) long. On uneven country the vehicles were generally run independently of each other. If the slope or condition of the ground so required, the two vehicles could be moved one after the other and thus make progress by successive stages.

Fig. 28. Saint Chamond chain-track motor mounting, electrically driven, with 194 mm. quick-firing gun, long type; fore-carriage, electrically driven, carrying the ammunition and the generating set for both vehicles.

The Saint-Chamond 280 mm. howitzer or mortar, carried on a chain-track mounting (fig. 29), though not so heavy as the 220 mm. (long type) Schneider gun and mounting, weighed with its tractor carrying the generating set and thirty rounds of ammunition about 56 tons in all, distributed almost equally over the two vehicles. The rough adjustment for direction the by the electric drive enabled a complete revolution to be made in 30 seconds (12 degrees per second); the fine adjustment effected by hand gear was at a rate of 3 seconds per degree. The petrol consumption was 2.2 gallons per mile.

Fig. 29. Saint Chamond chain-track motor mounting, with 280 mm mortar, type S, in firing position.

Although this type of arm has been so highly developed in France, it has practically neglected in England; in America, however, it has found favour, and no fewer than eight different models have been made and tested ranging from an anti-aircraft 75 mm. motor gun, with elevation up to 45 degrees, to a 240 mm. howitzer motor-propelled chain-track mounting having a total weight of 49.5 tons (Table II p.80).

Another vehicle which calls for special notice is the Christie combined wheel and chain-track gun carrier. The idea of constructing a vehicle that would be rapidly convertible from chain-track to wheels and *vice versa* had been worked out in France both for a lorry and a gun carrier by Colonel Rimailho, of Saint-Chamond. The Christie is arranged with four 36 in. solid rubber tyred wheels on each side. When used as a wheeled vehicle a chain-track is under the mudguard on each side ready for dropping on to the wheels, and the two centre pairs of wheels are raised off the ground, the front wheels acting as steering wheels in the usual way. As a chain-track vehicle, the centre pairs of wheels are lowered, the steering wheels are locked, and steering effected by clutches as on other chain-track machines. It is stated that this change can made in 15 minutes, and it is to this by improved design. The maximum speeds are 14.75 miles per hour wheeled, and 9 miles hour on the chain-track. The Christie carries a 155 mm. gun, 19.8 ft. long, and its total weight is 19.6 tons.

The essential features for chain-track machines.

1. *Gearing with the Ground.-*

The chain-link is usually formed with two projecting ribs which form the teeth that engage with the ground. Where the chain-track on the slab of earth between these teeth the horizontal shearing resistance of the earth may on grassland amount to 45 lb. sq. in., but as soon as the link begins to rise from the ground and the pressure the earth is relieved, the resistance falls rapidly. In clay ground, as soon as shearing commences, the interspaces the chain-links become filled with clay, and slipping will continue indefinitely. To avoid this, detachable spuds (called grousers in America) are used entering several inches into the ground. The resistance to forcing the normal surface through the ground may amount to 45 (or even 70) lb. per sq. in. of the vertical area transmitting the driving force. On soft or marshy ground it is the vertical pressure alone that need considered, the insistent weight being greatly reduced. As the class of land that requires this low insistent weight, the question of using spuds rarely arises.

2. *Insistent Weight.-*

It is necessary that war vehicles should be able to travel, not only where cavalry would be used, requiring a resistance of the ground of 30 lb., per sq. in., but also where infantry is concerned and the pressure on the foot of a man, about 9 lb. per sq. in. is the unit. Actual experience has shown that these vehicles should he capable of traversing even softer ground, as they cannot pick their way well as the infantryman, and in the later of small tanks it was reduced to about 6 lb. per sq. in. American commercial tractors have been made for working with an insistent weight of less than 2 lb. per sq. in.

The difficulties of traction of this kind increase greatly when it is required to cross snow. The ski-runner exerts a vertical pressure of only 0.75 lb. per sq. in., and the wearer of snow shoes about double this amount. A further difficulty arises with snow, which at temperatures near freezing point cakes on to the metal surfaces under pressure, and in the chain-track the ice so formed may cause jamming. At very low temperatures the snow remains loose and does not cake.

For soft ground it is possible to increase the surface by the use of overhung (non-symmetrical) chains, and it was necessary to fit some of the Saint-Chamond and many English tanks with chains thus widened.

3. *Power per Ton.-*

The power per ton varies in industrial tractors from 8 to 15 h.p. ; the agricultural tractor is usually capable of exerting a pull of from 35 to 60 per cent. of its weight. In this case extra weight is not carried on the tractor itself. The war machine, heavily armoured or loaded, is required to be capable of climbing slopes up to 45 degrees. The fact that in these vehicles the power transmitted to the track is in some cases over 80 per cent. of the weight enables this to be accomplished.

4. *Overall Dimensions.-*

The chain-track wears very rapidly if run on a hard road; it is therefore necessary to arrange for transport by rail. It can load itself if a ramp provided, each tank as it climbs running along the top of the platform waggons until it reaches its place in the train. It is necessary to arrange that any the tank, such as the gun-sponsons, that may project beyond the limits of the loading gauge, must be detachable or capable of being swung on pivots into the interior of the tank so as to give the necessary clearance. The height of the tank is limited not only by the loading gauge, but also by the necessity for reducing visibility; the height of tanks rarely exceeds 8.2 ft.

5. *Speed.-*

The slow speed of the commercial tractor varies from 1 to 3.5 miles per hour ; in the case of heavy lorries it ranges from 1.5 to 7.5 miles per hour. The British Whippet tanks were the first in which was recognised the principle laid down by Colonel Crompton that "real armouring is speed,"* and they had a range of 1.5 to 8.3 miles per hour.

* A similar expression was used by Lord Fisher, First Lord of the British Admiralty with regard to the armouring of battleships.

6. *'Effectiveness' or Immunity from Attracting Enemy Fire.-*

As a basis of comparison, Colonel Crompton laid down the principle that the 'effectiveness' of a tank, or its 'immunity factor,' is a function of the square of its speed divided by its height. This appears to have been largely justified by the changes of design that were made during the period over which the tanks were in active service.

7. *Control.-*

The steering, gear change, and brakes should all under the control of one man; this has been found necessary for safety and rapidity of manoeuvering with the ordinary automobile; it is even more necessary and, in fact, essential with the tank.

8. *Armouring.-*

The thickness of the armouring should vary, as on battleships according to the part of the tank to be protected and the angle which the surface is likely to present to enemy fire; a thickness of nearly 0.5 in. is required to stop rifle fire at very short range.

9. *Position of the Centre of Gravity.-*

This has a great influence on the capability of the tank to climb obstacles and to ascend and descend steep gradients. A position of the centre of gravity too high or too far back involved climbing an obstacle to a greater angle before equilibrium occurs and the tank rocks forward (see fig. 30). It must also be remembered that when the tank is running forward the effect of the couple due to the drive on the track is to produce a virtual displacement of the centre of gravity towards the rear of the vehicle (see fig. 31), and that for this reason also the centre of gravity should be kept forward.

These conditions, reversed when the tank is descending a hill on the brakes, require that the effective length of track in contact with ground should sufficient to prevent pitching. If the centre of the arc of contact with the ground be in front of the vertical through the virtual centre of gravity, the car will descend without pitching over, but if the centre of the arc falls on this line the tank will begin to turn over, and the brake resistance must reduced in order to diminish this tendency (fig. 32).

To reconcile this condition with that given for forward driving, it is necessary to keep the centre of gravity as low as possible and to design the curve of the chain to give requisite stability both in driving forward up the maximum climbable gradient and in braking in descending that gradient.

10. *The Curve of the Chain-Track in Contact with the Ground.-*

Not only must this meet the conditions requisite for running stability, but it must not involve so great a length of chain in contact with the ground as to make steering difficult. It has, moreover, a great influence on the variation in the insistent weight as the sinkage is increased. The earlier French chain-tracks were constructed on the Holt system with spring Suspension and a long bearing surface; the Renault tanks also spring suspension, but a shorter bearing surface. The English tanks were without springs, but had a very long bearing surface on the ground. The later Marks a more oval form of track, and although this form does not give the minimum sinkage, it has considerable advantage in making the steering more easy.

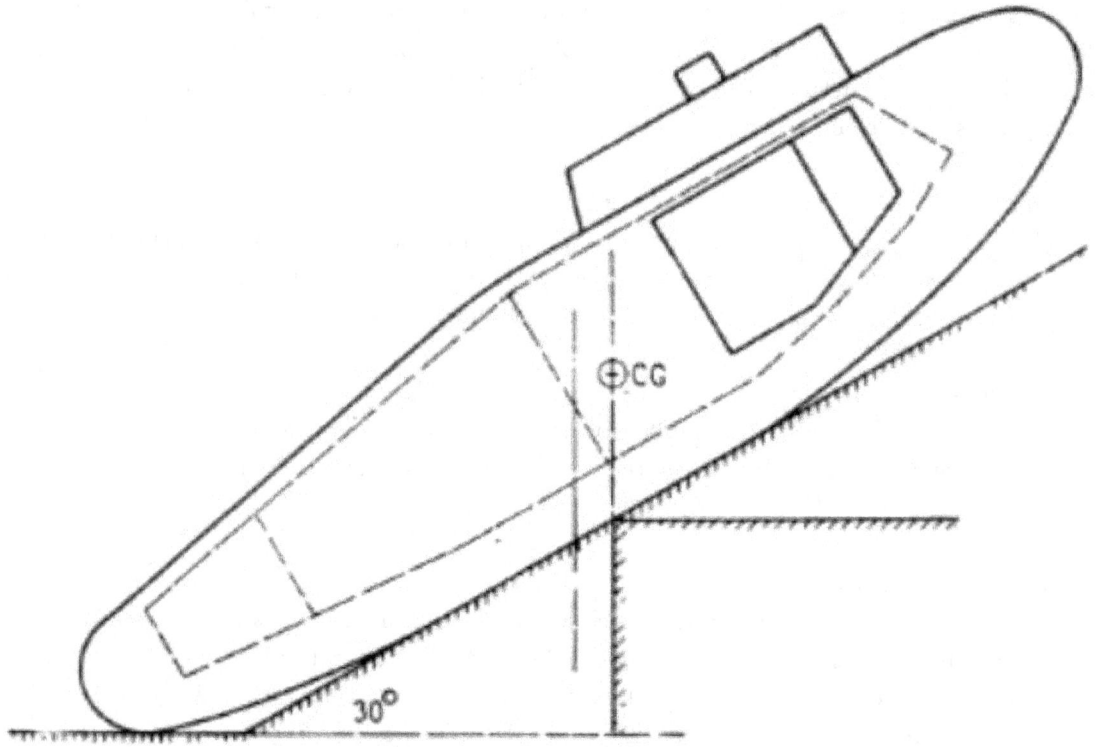

Fig. 30. Diagram showing conditions for stability when climbing gradient.

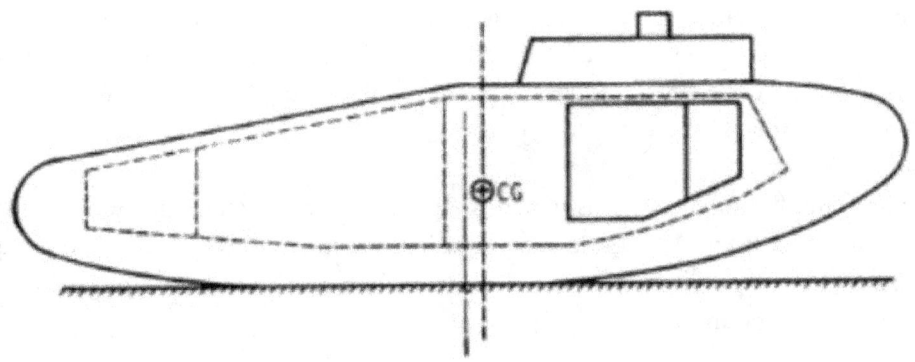

Fig. 31. Diagram showing virtual change of position of centre of gravity when running.

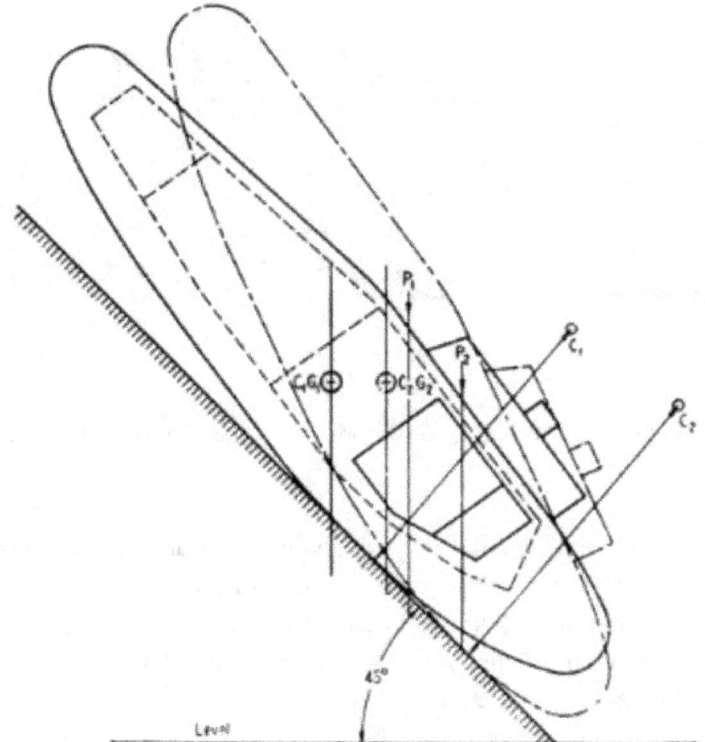

Fig. 32. Diagram showing brake action in descending a slope.

C1G1 – Centre of Gravity before tilting.
P1 – Vertical through virtual centre of gravity due to braking couple.
C1 – Centre of curvature of track at point of contact.
C2G2 – Vertical through virtual centre of gravity after tilting, brake still applied.
C2 – Centre of curvature of track at new point of contact.

11. *The Curve Traced by a Point on the Chain-Track.* –
The curve so described is worthy of study, as it shows the angle at which the descending portion of the chain in the front of the tank exercises pressure on an obstacle to be climbed. The behaviour of the tank when brought into contact with a fixed object such as a wall is also of great

interest. The after part of the chain remains in gear with the ground, provided this affords sufficient resistance, and the front of the vehicle climbs by reason of the frame of the vehicle moving over the track. The resultant of the shearing effort on the ground and of the weight of the tank tends to overturn the obstacle (fig. 33).*

* These points were illustrated by films, when the paper was read, and tanks were shown actually climbing up and down slopes of 45 degrees, crossing trenches and water, and also going through building. The gun-carrying chain-track motors were shown being driven into position and fired from the chain-track mountings.

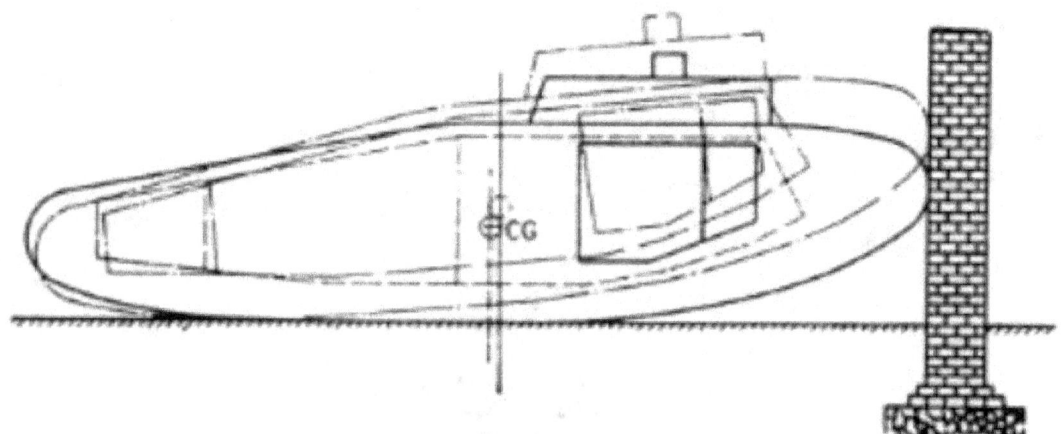

Fig. 33. Diagram showing conditions when pushing over an obstacle.

12. *Distance Run.-*

The distance actually run by tanks on roads and on the ground are widely different; according to the observations made by General Estienne during the war, out of a total average distance of 500 km. (310 miles) run by a tank, only 30 km. (19 miles) were run off the roads, and as running on the road out the chain-track much more than running over soft ground, it was necessary to transport the tanks as far as possible by rail, or, in the case of the lighter tanks, by wheel-tractors. Numerous trials and experiments have made in France, in England, and in America to fit the vehicle either with detachable wheels or with removeable chain-tracks. Up to the present no system has shown distinct superiority.

Possible improvements in the chain-track.

The hammering of hard - and particularly of iron-shod—wheels on the road has always been the greatest enemy to high speed in traction. Speed on the road has increased by the use of rubber tyres, and this has been still further increased by the substitution of the pneumatic for the solid tyre. Attempts have been made to obtain a more yielding contact, and before the war M. Kégresse Russia had produced a car capable of running on belt-tracks over loose, powdery snow. Trials made on a Citroën car fitted with the Kégresse attachment showed that it was to run on the belt-track roads, soft sand, steep sand banks, and even on loose, deep snow at a speed of over 20 miles per hour, or at three times the racing speed of a ski-runner on give and take snow.

It is obvious that the chain-track will find its future development to depend on more efficient springing than exists at present. If these improvements follow with the same rapidity as in the case of the automobile, snow as a covering to rough country will actually prove an aid to travel, and light cars may able to be run great distances in winter over country that would impracticable without its covering of snow.

It is probable that in the near future light cars will be seen capable of being run at will on the easier slopes of mountains, across fields, over the snows of the Arctic and the sands of the desert, as well as on the ordinary road, for, as General Estienne has said (trans.): *"The appearance the battlefield of chain-track mechanical vehicles was an event the importance of which equals that of the invention of gun powder."*

The use of gunpowder in cannon was the first step towards the adoption of explosives for blasting and tunnelling; it is possible that the use of the chain-track on war machines may lead to equally important developments in the tractors used for other and peaceful purposes.

Some idea of the magnitude assumed by mechanical warfare may be obtained from the fact that amongst the allied nations over 28,400 tanks were ordered of which over 7,100 were actually constructed.

The value of tanks made and order amounted to about 1,500,000,000 francs, of £60,000,000, and placed side by side they would have covered a width of about 60 kilometres, or 40 miles.

In his paper the author has endeavoured to pay just tribute to the engineers whose fertile brains have conceived and elaborated the details, the creators of the tanks, and the adaptors of the chain-track to carrying artillery - and to express the admiration and thanks of all, particularly to Colonel R. E. B. Crompton, of England; to Mr. William Strait, of the United States; to General Estienne, M. E. Brillié, and to Colonel Rimailho, of France, with the assurance of our profound gratitude.

Before concluding, the author wishes to express his sincere thanks to the British Army Council, to the French Minister of War, and to the American Chief of Ordnance, as well as to the Imperial War Museum and the Institution of Mechanical Engineers, for the permission which these several Governments and institutions have been kind enough to give him for the reproduction in lantern slides, illustrations, and films of the different machines to which reference has been made.

The author wishes also to record his gratitude to M. Brillié, of the firm of Schneider et Cie.; to Colonel Rimailho and M. Berthier. of the Forges et Aciéries de la Marine et d'Homécourt; to M. Jannin, of the Renault Works; to the management of the Fiat Works of Turin; and also to the devoted secretary of the Société des Ingénieurs de France, M, de Dax, for the assistance and collaboration which they have all so generously given to him in the presentation of this paper.

www.ingramcontent.com/pod-product-compliance
Lightning Source LLC
Chambersburg PA
CBHW060411220526
45465CB00008B/2847